ALEKSANDRA THOMSON
RACHAEL GOOBERMAN-HILL

NURTURING EQUALITY, DIVERSITY AND INCLUSION

Support for Research Careers in Health and Biomedicine

T0051446

POLICY PRESS SHORTS POLICY & PRACTICE

First published in Great Britain in 2024 by

Policy Press, an imprint of
Bristol University Press
University of Bristol
1–9 Old Park Hill
Bristol
BS2 8BB
UK
t: +44 (0)117 374 6645
e: bup-info@bristol.ac.uk

Details of international sales and distribution partners are available at
policy.bristoluniversitypress.co.uk

British Library Cataloguing in Publication Data
A catalogue record for this book is available from the British Library

ISBN 978-1-4473-6799-4 paperback
ISBN 978-1-4473-6800-7 ePub
ISBN 978-1-4473-6801-4 ePdf

Cover design: David Worth
Front cover image: istockphoto/tawanlubfah

Contents

List of tables, figures and boxes

Tables

Figures

Boxes

List of abbreviations

ADHD	attention deficit hyperactivity disorder
AGCAS	Association of Graduate Careers Advisory Services
APPG	All-Party Parliamentary Group
AR2O	achievement relative to opportunity
Athena SWAN	Athena Scientific Women's Academic Network
BAME	Black, Asian, and Minority Ethnic (see Chapter 1 for discussion about this abbreviation)
BEIS	Department for Business, Energy & Industrial Strategy
BME	Black and Minority Ethnic
CoP	community of practice
CRAC	Careers Research & Advisory Centre (includes Vitae)
CV	curriculum vitae
DAISY	Diversity and Inclusion Survey developed by EDIS
ECR	early career researcher
EDI	equality, diversity and inclusion
EDIS	equality, diversity and inclusion in science and health
EHRC	Equality and Human Rights Commission
EIGE	European Institute for Gender Equality
EU	European Union

FNR	National Research Fund, Luxembourg
FORGEN	Funding Organisations for Gender CoP
GATI	Gender Advancement for Transforming Institutions
HE	higher education
HEI	higher education institution
HESA	Higher Education Statistics Agency
HR	human resources
IAT	implicit association tests
ICF	International Classification of Functioning, Disability and Health
LGBTQIA+	lesbian, gay, bisexual, transgender, queer, intersex, asexual, and more (see Chapter 1 for discussion about this abbreviation)
NCSES	National Center for Science and Engineering Statistics
NCV	narrative curriculum vitae
NHS	National Health Service
NIHR	National Institute for Health and Care Research
NUS	National Union of Students
OECD	Organisation for Economic Co-operation and Development
PhD	Doctor of Philosophy
PSED	public sector equality duty
R4R	résumé for researchers
R&D	research and development
REC	Race Equality Charter
REF	Research Excellence Framework
RFBR	Regional Foundation for Biomedical Research
SAGE	Science in Australia Gender Equity
SciCV	scientific curriculum vitae
SEA Change	STEM equity achievement change
SF DORA	San Francisco Declaration on Research Assessment

SFI	Science Foundation Ireland
SIRG	starting investigator research grant
SME	small and medium enterprise
SNSF	Swiss National Science Foundation
STEM	science, technology, engineering, mathematics
STEMM	science, technology, engineering, mathematics, medicine
UCL	University College London
UK	United Kingdom
UKRI	United Kingdom Research and Innovation
UN	United Nations
US	United States of America
UUK	Universities United Kingdom
WHO	World Health Organization
WT	Wellcome Trust

About the authors

It is important to acknowledge our positionality as authors and the subjectivities brought into the book. We recognise that knowledge is always situated knowledge, and that our own context, experiences and perspectives shape knowledge that we produce. Therefore, we wish to locate ourselves in our work contexts:

Aleksandra (Ola) Thomson is Lecturer in Business and Management at Bournemouth University Business School, UK. Before this role, she worked as Research Associate in Equality, Diversity and Inclusion at the Elizabeth Blackwell Institute for Health Research at the University of Bristol, and at Loughborough University Business School in a Horizon 2020 project focusing on gender equality in research and academia. She grew up in Poland and completed her postgraduate qualification in the UK. Ola's interest in gender equality in knowledge work and academic and research institutions has expanded to considerations of intersectionality of compounding social categories. Her focus is to advance understanding of how knowledge workers, academics and researchers can be practically helped to thrive in their careers, and how equality, diversity and inclusion can be nurtured by individuals, EDI practitioners and institutions.

Rachael Gooberman-Hill is Professor of Health and Anthropology and Executive Director of the Elizabeth

Blackwell Institute for Health Research at the University of Bristol, UK. She was trained as a social anthropologist in Scotland, having grown up in the north-west England and subsequently in Ireland. Rachael is now based in south-west England and is a mother and full-time permanent employee at the University of Bristol. She has worked in interdisciplinary health research for several years, valuing the chance to work with individuals and groups from many backgrounds. Her work in health has included a focus on health conditions that affect people in myriad and often serious ways—for instance, including work on the impact of, and treatment for, long-term pain after surgery. Rachael is interested in how the research system, as a whole, can attend to the diversity of experience and thought, so that research is relevant and can make a difference in people's lives, including for members of the research community and people who ultimately benefit from research.

Our roles in the book were different and complementary. We both conceptualised the work and designed it together. The first author conducted the study interviews and was largely responsible for drawing the existing literature together to produce the initial draft text. We then worked together to shape the structure and interpretations.

Acknowledgements

We wrote this book as part of our roles at the Elizabeth Blackwell Institute for Health Research at the University of Bristol, UK. As such, the work was supported by the Elizabeth Blackwell Institute and the Wellcome Trust Institutional Strategic Support Fund, grant number 204813/Z/16/Z. Our thanks to the Wellcome Trust for this vital funding. Thank you to the eight interview participants who generously gave their time and expertise: we are mindful that sharing experiences relating to equality, diversity and inclusion takes particular energy and effort.

We have benefited immensely from working with the whole Elizabeth Blackwell Institute team, especially Dr Nina Couzin, as well as from working with and listening to colleagues across the wider research ecosystem. This includes the readers and the anonymous reviewers who provided such helpful and generous feedback on our original proposal to Policy Press and the book draft. Our ideas have been shaped by conversations with many people over many years, and our thinking will always be 'work in progress'. Thanks are due to everyone for these conversations: you are numerous, and we hope that you know who you are and that this short book feels like a further step in the right direction. Finally, we hope that the content resonates with experience and is of use in practice.

Introduction

Attention to equality, diversity and inclusion (EDI) is growing across all research fields and disciplines. This book is largely focused on UK health and biomedical research, also drawing on international and interdisciplinary evidence to bring together examples of key concepts, current thinking and emerging approaches.

Key to the research ecosystem is the body of researchers, who all bring their own commitment, dedication, experiences, circumstances, views and interests. These backgrounds can inform the paths that they take and the focus of their research, including with particular groups, populations, contexts and health conditions. Indeed, researchers interviewed as part of the work to develop this book became EDI advocates and engaged with EDI topics for a variety of reasons, including their own experiences and backgrounds. Such personal motivation has considerable benefits to EDI within the research community and to the design of research itself. Researchers are often advocates for inclusion, want to make a difference in the world and act as key role models for future generations, for instance through their teaching and mentorship. The research ecosystem also depends on the work of researchers through their formulation of research ideas, conduct of peer review, and communication of research to their peers and to members of the public.

The book focuses on researchers and research careers, but we acknowledge that attention to research design, analysis, pedagogic content, science communication and public engagement provides distinct, yet complementary, foci. Although EDI within research is often framed with attention to staff and student bodies, EDI considerations apply equally to stocks of knowledge, ideas and philosophies. These include dominant paradigms in curricula and calls for decolonisation of knowledge, application of sex, gender, race, ethnicity, disability, neurodivergence, and intersectional analyses in research studies (see, for instance, Bentley et al, 2017; Schiebinger et al, 2020; Botha, 2021; Wong et al, 2021; Yerbury and Yerbury, 2021). There are several questions to address in research itself: How do we devise research questions and collect data? What tools and ideas do we apply that may perpetuate inequalities? Why are some questions more important than others? How do we decide which communities our research serves? EDI also matters when thinking about how best to communicate research so that findings are available to all. This is an international concern. For instance, writing about responsible science communication in Africa, Elizabeth Rasekoala underscores its key role in countering misinformation and the need to 'make way for new diverse and accessible narratives that speak to a wider knowledge base and resonate with the lived experiences of all in society' (Rasekoala, 2022, p. 5). Design, conduct and communication of research are important fields of inquiry—questions that deserve separate and undivided attention—and we recognise that we are unable to attend to these topics in this short book.

This book suggests that there is a need for institutional change and that this journey can be supported through carefully designed interventions. These are needed to equalise and democratise access, opportunities and outcomes for all researchers regardless of background, identity or circumstances. To understand the effectiveness of interventions for change, we draw on qualitative and quantitative material from existing literature. The material that we have brought together in this book is a curated product of our work to collate and synthesise material garnered through literature searches, our knowledge of outputs and studies by various EDI stakeholders and equality networks. This means that we draw on academic and grey literature that includes both quantitative and qualitative material. The literature is complemented by qualitative material from narrative interviews conducted in 2022, and we provide detailed information about the study interviews in the appendix. The interviews provide information about experiences, including those relating to interventions designed to address EDI. Sometimes simply asking people how something was for them provides information about how a well-intentioned approach makes a positive difference or about any unintended consequences, which in EDI might include entrenchment of inequalities.

It is important to acknowledge the pace at which evidence is moving in the field of EDI. Recent developments reflected in the book include reviews commissioned by UK Research and Innovation (UKRI), which sought to collate national and international evidence about what works to address EDI challenges (Guyan and Douglas Oloyede, 2019; Moody and Aldercotte, 2019). More recent reports continue to build knowledge about the scale and shape of equality matters in health and biomedical sciences workforces. For instance, in 2022, the UK's Biotechnology and Biological Sciences Research Council (BBSRC) within UKRI published a report conducted by the University of York into inequality in early careers in the Life Sciences. That report provides a blueprint for empirical investigations in further fields and disciplines (Dias Lopes and Wakeling, 2022). No doubt in the period between completion of our book in January 2023 and its publication, more useful evidence will have been produced that will help to underpin approaches to successful interventions in EDI. Therefore, we encourage the reader to continually consult the latest available evidence.

Notes on timing, content and language

This book is based on work to bring together existing and recent knowledge and experience of EDI in health and biomedical research careers. Over two years, the authors sought to understand and explore literature and the current situation in research careers, complemented by interviews with individuals active in EDI. Therefore, the project necessarily reflects a particular time and set of perspectives and does not include every possibility. The work started during the 2021 pandemic and ended in January 2023 as the UK and Europe were exiting acute pandemic situations but were impacted by a new geopolitical crisis in Europe, which we acknowledge as one of many ongoing crises globally.

The nature of EDI work and its diverse foci, themes, identities and contexts, and the limits of a short-format publication, mean that some literature has been included in the book, and some has inevitably been left out. As secondary research findings have been included in the book, there might be some inevitable omissions, although we have tried to be as accurate as possible in our representation of the work of others. Our aim was to provide a piece of work that is of value, freely available and useful to readers and we hope that the book does so.

Finally, we explore some terminology and definitions in Chapter 1. Language changes over time, has deep meaning and impact. We are mindful that language can replicate and reinforce inequalities. In this book we follow Bristol University Press's style guide and capitalise 'White' and 'Black' where used, but respect the discussions about capitalisation of these terms, including relative capitalisation and the need to recognise long-standing discrimination.

ONE

Equality, diversity and inclusion: key definitions

This chapter defines and discusses key concepts and terminology relating to health and biomedical research, equality, diversity and inclusion (EDI), disability, neurodiversity, race and ethnicity, sex and gender, and sexual orientation. These terms evolve, are debated, and can be the subject of much contestation. The chapter will acknowledge additional categories of diversity that might be impacting on researchers' careers, and the concept of intersectionality will be introduced. Concepts that are introduced will be referred to and revisited in subsequent chapters.

What is health and biomedical research?

Research work is creative, systematic effort that increases the 'stock of knowledge' (OECD, 2015). In the fields of health and biomedicine, research has great potential to benefit humanity. Such research takes place in myriad contexts and depends on the great skill and dedication of the research workforce. Health and biomedical research delivers answers that help to protect and maintain health, provides knowledge that can help to treat illnesses, and provides information that points the way to future solutions. In recent years, advances fostered by these areas of research are numerous, including better understanding of causes of illness, development of new treatments and preventive measures, as well as knowledge that can help to support change that impacts on individual and societal health and wellbeing. Both 'positive' and 'negative' or 'null' results are important to knowledge advancement. Without research, there would be no treatments for tuberculosis, no vaccines for COVID-19, and no progress in understanding the complex interplay between social structures and health outcomes.

Health and biomedical research is conducted by people with many disciplinary backgrounds, including natural, social and behavioural sciences, as well as arts, humanities and engineering. The ecosystem is rich in skills, with technical, professional and clinical experts working in close collaboration and teams. Health and biomedical research may have single-discipline or interdisciplinary flavour and team approaches to research are the norm. In this book, the term 'health and biomedical research' is used to refer to research that seeks to understand matters relating to human health: this is necessarily broad but also focused, with disciplines varied and often working together in interdisciplinary ways.

A considerable body of health and biomedical research focuses on reducing inequalities in health, including through understanding disparate health outcomes, improving health and enhancing fairer access to healthcare. This focus is echoed by work to improve fairness in the system that creates knowledge, including in relation to career development and support.

Research takes place in a complex ecosystem, comprising public organisations as well as the third sector (such as in charities), and private sector organisations ranging from small and medium enterprises (SMEs) to large, international corporations. In the UK alone, the life sciences industry employed over a quarter of a million people in 2020 (BEIS,

2022). Higher education institutions (HEIs) are particularly visible although not the largest sector in the UK research ecosystem. In 2020–21, there were 170 HEIs in the UK (UUK, 2022): all provide education and the majority conduct research of immense variety, with quality assessed through the national Research Excellence Framework (REF) exercise. In 2021, 157 HEIs made submissions to REF, of which 129 submitted their research activity to Panel A, which consisted of Clinical Medicine; Public Health, Health Services and Primary Care; Allied Health Professions, Dentistry, Nursing and Pharmacy; Psychology, Psychiatry and Neuroscience; Biological Sciences; and Agriculture, Food and Veterinary Sciences (UKRI, 2022a). Research into health-related areas was also submitted into other panels. The UK's strength in biomedical sciences is also demonstrated and boosted through research centres, for example the Francis Crick Institute and the Medical Research Council's Laboratory of Molecular Biology, both of which receive public funding alongside other support such as from charities. Private companies, including pharmaceutical firms, also make major important contribution to UK research and science.

What is equality, diversity and inclusion?

EDI in research careers has been subject of increasing attention, especially during and in the aftermath of the COVID-19 pandemic. Career inequalities are not unique to health and biomedical research, but reflect broader issues persisting in wider academic and research and innovation settings, that have historically disadvantaged women and individuals from minority ethnic backgrounds. Seeking to redress the lack of equality, enhance diversity and inclusion will nurture science, its applicability and relevance, as well as support a dignified, respectful research environment. This has never been more vital. Focus on equalities is part of wider attention to research culture that places people at the heart of all research endeavours.

Discussions of EDI are not new, although priorities have changed over time and the lexicon relating to EDI has evolved. Across the ecosystem research, current initiatives that address EDI include activities focused on equal opportunities, anti-racism, gender equality, or employee wellbeing. The use of 'equality, diversity and inclusion' to encompass this variety provides a common language and frame of reference. Despite their ubiquity, the terms are not always clear and, sometimes, their meanings and use are contested. These terms will be presented in turn.

Equality and equity

'Equality' and 'equity' have related but different meanings. 'Equality' refers to provision of fair and same treatment to everyone. 'Equity' usually refers to recognition of the context of disadvantage that people may face, including in relation to individual characteristics and social structures and systems in which they live. People may need to be provided with different resources or support in accordance with individual, structural and social circumstances to nurture equal access to opportunities and equal outcomes.

Although equality and equity are often understood in these ways, organisations may use the terms in different and specific ways. For instance, UKRI explains that equality is: 'treating *everyone* the same and giving *everyone* access to the same opportunities', and equity is: 'creating fair access, opportunity and advancement for people from *under-represented groups*' (emphasis by the authors, UKRI, 2022b). In the context of gender, the European Institute for Gender Equality (EIGE) defines equality as 'equal rights, responsibilities and opportunities of women and men and girls and boys'; and equity as 'provision of fairness and justice in the distribution of benefits and responsibilities between women and men' (EIGE, 2016b, np). The focus shifts to distributive fairness in the equity definition.

Equality and equity have different meanings but are often used interchangeably or even combined. In the context of academia and gender, Lotte Bailyn (2003) argues that conflation of equity with equality stems from assumptions that the academic workplace exists in a vacuum and is entirely independent from other spheres of life. This separation disregards the distinct life experiences of men and women who have different opportunities to succeed. The academic 'ideal' worker norm is often reported to be more challenging for women to achieve, because they tend to hold more domestic responsibilities than men. As such, equal opportunity, even if it exists, is never equitable as people's backgrounds, experiences and different forms of human capital vary depending on their circumstances and context, who we are, where we are, and where we come from. To address variation between people some tailored mechanisms can be provided, such as access to training, opportunities, networks, resources, mentoring and other support (Putnam-Walkerly and Russell, 2016). However, these interventions will have limited impact unless systemic and structural social injustices that occur way before and during employment and career development can be addressed. These injustices also happen even after careers have been developed or completed, for example, in the unequal pension arrangements and income received by men, women and minoritised groups.

The recent shift of focus from equality to equity is illustrated in increasing use of the terms 'diversity, equity and inclusion': a change in word order as well as replacement of 'equality' with 'equity', particularly in the US. However, this change has generated some challenges. For instance, the change from equality to equity may de-emphasise the attention paid to regulation and legislation. Rather than seeking to bring an end to systemic issues that contribute to inequality, equity perspectives may necessitate differentiating between people and making decisions about their presupposed needs according to certain protected characteristics. This focus on particular groups of individuals and compensatory programmes may inhibit or block opportunities for deeper structural changes to address the root causes of inequalities in the first place (Ryan, 1971). Such perspectives are described as the 'deficit' approach (Valencia, 2010). An equity perspective may therefore overemphasise difference and diversity categories, which could create unintended and broader consequences by bolstering stereotypes and essentialising; for example, by emphasising that there is, 'an innate female nature, an essence of femininity, or innate racial characteristics that can explain social relationships such as those of gender and race' (Ramazanoglu and Holland, 2002, p. 58). Equity perspectives can also create risks of diffusion of collective interests into individual-level concerns, diluting the focus on systems of inequality.

Noting the debate and difference between the terms 'equity' and 'equality', the original and widely understood phrase term of 'equality, diversity and inclusion' will be used in this book, using the acronym 'EDI' for reading ease.

Diversity

'Diversity' usually refers to any one or more strands of difference. These include social or cultural groups, backgrounds and various aspects of identity. For instance, diversity is often referred to in relation to visions of workplaces, with diverse work environments characterised by wide ranges of people representing categories of difference in relation to gender, age, race, ethnicity, sexual orientation, (dis)ability, culture, nationality, religion, socio-economic background, class, education, neurodiversity, among further aspects of difference.

Some commentators suggest that 'diversity'—a relatively new but widely used term—originated in US managerial discourse (for discussion see Ahmed, 2012), and seek to understand why this term has replaced previous descriptors of inequality. One of the suggested explanations is related to a focus on positive visions of workplaces and positive language, but at the same time displacement of the language of 'equal opportunities' or 'anti-racism'. Sara Ahmed (2012) argues that this reframing is problematic, not only because it is constructed to be more 'palatable' and acts as a corporate aesthetic of branding. On a more subliminal level, it may also act as a distraction from efforts towards a deeper institutional change, leaving organisations to concentrate efforts into numerical and thus visual diversity, which by default will not produce inclusive work environments.

Preferences for concepts and language may also relate to norms in certain research contexts in which neutrality and objectivity are key values, and this is especially valid in health and biomedicine. The recent linguistic transition to 'diversity' might indicate the existence of some discomfort with 'emotive and emancipatory' connotations of 'equal opportunities' (Özbilgin, 2009, p. 5). At the same time, diversity has been increasingly employed as a vehicle to attract talent; this can be seen in staff and student recruitment approaches and accords with positive attributions of the language of diversity. Typing 'diversity' into an online search engine generates a plethora of pleasing graphics, including colourful letters, puzzles, hands, all of which signal that diversity is happy, harmonious and desirable.

Inclusion

In theory, diversity can be nurtured through inclusion, which means doing something positive 'with' or 'to' diversity. The act of inclusion can take many forms, is context-dependent, and can take place at organisational and individual levels. Often, inclusivity in workplaces is taken to mean the creation of welcoming and respectful environments in which difference is celebrated, everyone feels valued and wanted.

As well as comprising action, inclusion is a feeling conceptualised as 'the degree to which individuals feel a part of critical organizational processes such as access to information, connectedness to co-workers, and ability to participate in and influence the decision-making process' (Mor Barak et al, 2001, p. 73). Similarly, but with emphasis on the actions of others, inclusion is also 'the degree to which an employee is accepted and treated as an insider by others in a work system' (Pelled et al, 1999, p. 1014). Feeling included can manifest as a sense of belonging.

The opposite of inclusion is exclusion, which is also an action or a feeling and has particularly strong moral overtones: seldom do people say that they like feeling excluded. But both inclusion and exclusion can be experienced as something done to a person or group, either by another individual, an environment, structure or process. The fact that inclusion matters, relates to differences in power, influence and also to access to resources. Inclusion takes place in societies and organisations, and the hierarchies within and between them.

Hierarchy and power can create dynamics in which individuals and groups who are excluded may not be in a position to make changes that enable inclusion. To achieve institutional change, allies will play a critical role in advocating for groups who are excluded (see Chapter 4 for 'allyship' and 'communities of practice'). Inclusion is not only about bringing people in, but is about 'making strangers into subjects, those who in being included are also willing to consent to the terms of inclusion' (Ahmed, 2012, p. 163). Recognising the importance of this shift runs through interventions that prioritise the group of people these acts purport to help, underpinning the adage of 'nothing about us without us' where co-production and co-evaluation of EDI interventions are considered at every stage of design, conceptualisation and delivery of any approaches that address inclusion. In the design and development of initiatives, there is a need to listen to people and groups who are generous to share their lived experience of exclusion and disadvantage, but without placing additional burdens. For instance, this may include attending to the experiences of people whose projects are not granted funding or whose career trajectories are not aligned with notions of disciplinary norms.

Defining categories of diversity

There are many areas of diversity, which can be described as categories that relate to individuals' characteristics. Categories have some internal cohesion. Defining categories can be helpful, although doing so is not intended to essentialise groups of people, nor to oversimplify the humanity of diversity, nor to underplay the importance of intersectional approaches, in which the overlap and interplay between categories are considered.

In the UK, the Equality Act 2010 protects nine specific characteristics: age, disability, gender reassignment, marriage and civil partnership, pregnancy and maternity, race, religion or belief, sex, and sexual orientation. Race includes colour, nationality, citizenship, ethnic or national origins. These are important categories that provide a basis for thinking about diversity in careers. This section defines and discusses the areas of diversity that have been most thoroughly explored in literature to date: these are sex and gender, sexual orientation, race and ethnicity, disability, neurodiversity, and the overarching and amplifying impact of intersectionality. These categories are presented in alphabetical order, followed by a section addressing intersectionality.

In addition, emerging work highlights the importance of further categories that shape research careers. Therefore, this section also touches upon diversity dimensions that can guide approaches to issues of equality and inclusion.

Disability

Globally, 15 per cent of the population (around one billion people) live with disability (WHO, 2011). In the UK, around one in five people report a disability and more than 4.8 million disabled people are in work (Office for National Statistics, 2022). The employment rate of disabled people is around 54 per cent, compared with just over 80 per cent of people without disabilities. The known percentage of all disabled academic staff in the UK (those who had declared) showed that they were underrepresented at just over 5 per cent in 2020–21 (HESA, 2022a). There are generally more students than academic staff with a known disability, and this is also the case among biomedical doctoral graduates at 9 per cent (CRAC, 2022).

The UK's Equality Act 2010 defines disability as a 'physical or mental impairment that has a "substantial" and "long-term" negative effect on your ability to do normal daily activities'.[1] Rather than focusing on specific disabilities that are legally defined, the legislation prioritises the impact that a person experiences. Scholars have discussed definitions of disability at length and often with attention to inclusion and dignity that some of the language and approaches to disability can bolster or undermine. There has been a move away from a medical approach to disability that focuses on individuals to a more holistic approach that recognises the role of social and environmental factors on disability. This shift in thinking is reflected in the WHO's International Classification of Functioning, Disability and Health (ICF), which now encompasses the influence of social and environmental factors and defines disability as: 'an umbrella term for impairments, activity limitations and participation restrictions. It denotes the negative aspects of the interaction between an individual (with a health condition) and that individual's contextual factors (environmental and personal factors)' (WHO, 2001, p. 221).

In contrast, civil society organisations, such as the National Association of Disabled Staff Networks, endorse and support the social model of disability (NADSN, 2021). The social model suggests that it is the environment or society that create a disabling effect: a person in a wheelchair is disabled not by their inability to walk, but inadequate infrastructure to move easily, such as the absence of ramps, lifts, and accessible facilities (Oliver, 1983). Mike Oliver's definition of disability proposed that disadvantages or restrictions result from 'the political, economic, and cultural norms of a society which takes little or no account of people who have impairments and thus excludes them from mainstream activity. (Therefore disability, like racism or sexism, is discrimination and social oppression)' (1983, p. 16). As such, in contrast to the preferred term 'people with disabilities' prevalent in the US, organisations that conceptualise disability in line with the social model (especially in the UK) prefer to use the term 'disabled people' to purposefully emphasise the disabling effects of ableist social and structural assumptions and expectations (Lawson, 2022).

The social model alone is not sufficient to understand, explain, and address the experiences of disabled people (Oliver, 2013). For example, Anna Lawson and Angharad Beckett emphasise the human rights model, which focuses on the 'inherent dignity of the human being' as complementary to the social model (2021, p. 349). This model transcends the issue of belonging as a political endeavour to 'belonging to the human race', and it emphasises disability policy (Lawson and Beckett, 2021, p. 368) that is critical to systemic change. However, it is the social model that has gained traction within the context of EDI in the workplace (Oliver, 2013). This is because EDI efforts have focused to increase fairness and equalise opportunities for career prospects, promotion and opportunities. However, as argued by Oliver, even though the social model has helped to identify many of the disabling barriers in the workplace, 'the solutions offered have usually been based on an individual model of disability' (p. 1025), rather than approaching the issues as a systemic and institutional problem to tackle.

Nevertheless, individual disclosure of disability to employers is important to underpin work that can help disabled researchers to be included and feel enabled to thrive in research and academic settings. Disclosure of disability can signal people's perception that they will not be disadvantaged in relation to their disability. This requires faith in the fairness of the system and so rates of disclosure can serve as a kind of litmus test for inclusive culture. Open and transparent disclosure of disability brings opportunities for EDI practitioners to investigate rather than assume what barriers people face, what everyday impacts of these barriers are, and how they can be minimised or removed.

Disabled people report experiences of different forms of discrimination in the workplace. This can include limited access to development opportunities and being subjected to offensive behaviour (Foster and Scott, 2015). It has also been shown that disabled researchers are not nurtured to thrive in the highly competitive environment that can characterise some science careers. Typical activities that are undertaken by researchers, such as applying for external research funding, travel to and presentation at conferences, networking and attending events can be problematic for disabled researchers if the appropriate structures and processes are not in place. Even though disability adjustments can be requested and secured, individuals may encounter unhelpful and unnecessary barriers to modifications and institutional support (see Chapter 3 on 'institutional ableism'). For individuals, whose condition cannot be unequivocally defined as disability, but who are chronically ill or neurodiverse, self-identification through disclosure and adjustments may be therefore even more challenging to access (Yerbury and Yerbury, 2021). Moreover, individuals who are neurodivergent may find themselves undergoing a disciplinary action or performance management as a result of managers' lack of awareness of their condition and the extent of its impact on work (CRAC, 2020).

Neurodiversity

Neurodiversity includes Autism, Attention Deficit Hyperactivity Disorder (ADHD), Dyspraxia, Dyslexia, and a range of other variations. Although not necessarily understood as disabilities, in terms of legal definitions such diversities often sit under the protections afforded to disabilities. The term neurodiversity was first proposed by Australian sociologist Judy Singer (1998) to challenge the view that such divergence should be seen through the lens of diagnosis and to focus instead on matters relating to equality. This is not to say that diagnosis is not helpful, and neurodiversity is not often approached with attention to inclusion. For instance, in relation to learning disability, approaches known as 'de-differentiation' seek to include people with learning disabilities within broader groups of disabled people, or wider society (Banks et al, 2020). Neurodiversity can be seen as an identity characteristic that makes individuals unique, equipped with distinctive skills and abilities, and 'an equally valid part of human diversity' (Kapp et al, 2013, p. 68). As such, neurodiversity, in the same way as diversity, necessitates efforts to promote equality and inclusion of 'neurological minorities' (Baumer and Fruch, 2021) to boost these individuals' opportunities to thrive.

In terms of careers and lived experience, neurodiversity varies greatly from person to person. In the social and physical worlds of workplaces, perceived stigma, and a lack of adjustments to work type and infrastructure can create substantial barriers to neurodivergent researchers. There is also the impact of neurodivergence after work has finished. For instance, studies highlight the exhaustion and the consequences on sense of authenticity experienced by autistic people in social situations, who spent their time camouflaging to mask or compensate for their neurodivergence (Hull et al, 2017). This has very real repercussions, as autistic students are ten times more likely to drop out of university than their non-autistic peers (Gurbuz et al, 2019). Therefore, communication and mutual understanding are key to nurturing autistic researchers:

> Individuals can coexist if we all—neurotypical and neurodiverse individuals—in clear minds, make a connection in a calm way with each other by communicating in a collaborative way, in the spirit of cooperation, whereby both individuals understand each

other, with the correct content in context. (Larkworthy, 2022, np; also Gurnett and Morton, 2021; Act for Autism, 2022)

Neurodivergent researchers' experiences play an important role in mainstreaming neurodiversity in research content, improving knowledge of barriers, challenges, opportunities, and strengths in relation to these conditions. Recently, a growing community of autistic scientists have come together to study neurodivergence, share opportunities, make significant contributions to knowledge about autism through publishing, and serve as editors and reviewers of autism journals (Nuwer, 2020). Still, accounts of the lived experience of autistic researchers illuminating the complexities in navigating academic terrain are rare (for a notable exception in the field of psychology, see Botha, 2021). Furthermore, underrepresentation of individuals at the intersections of race, ethnicity and social class in autism research has contributed to health and healthcare inequities (Maye et al, 2021). Nurturing the careers of autistic researchers is thus crucial for widening participation in autism research.

Race and ethnicity

Race and ethnicity, and associated terminology, are used in myriad ways according to discipline and perspective. In health and biomedical research, the term 'race' is often understood through the lens of biology. This use is particularly apparent in the long-standing concern about inadequate diversity in research and in efforts to change this. For example, genomic studies benefit from attention to diversity in their design, as this improves our understanding of disease, including detection, diagnosis, drug development and design of clinical care (Fatumo et al, 2022). However, caution must be exercised to avoid viewing racial groups as 'distinct homogenous blocks' (Lim et al, 2021, p. 131). This is because more genetic differences can be found within each race than between races. As such, race is appropriately viewed as a social rather than fixed or solely biological concept if bias is to be eliminated in research and therefore in evidence-based healthcare (Lim et al, 2021).

Social sciences have a long history of understanding ethnicity as a sense of belonging and identity. A useful and current definition of ethnicity relates to group identity, and ethnic group as a 'collection of people or a social group whose members identify with each other through a common heritage, consisting of a common culture that may also include a shared language or dialect [but also] common ancestry, descent, diet, religion or race' (Darko, 2021, p. 9).

Sometimes race and ethnicity are conflated in ways that may not be helpful. Folúké Adébísí considers the use of the word 'ethnicity' to replace 'race' as problematic and obfuscating. This is because such replacement implies that only non-White people are ethnic, illuminating the power White people possess to make their race invisible (Adébísí, 2019). However, everyone has a specific ethnicity, and this is not something unique to minorities. The hybrid term race/ethnicity is unhelpful, as 'the concept of race is vital to understanding racism, whether historical or contemporary, and in referring to legislation and policy to reverse the effects of racism' (Johnson et al, 2019, p. 87).

Another example of unhelpful consolidation is the use of BAME (or BME) as a widely used acronym for Black, Asian, and Minority Ethnic or Black and Minority Ethnic, respectively. At the time of writing (2022) a topic field search in the Web of Science Core Collection for the two terms suggested that in the last decade there has been a nine-fold increase in their use. However, the popularity of the acronyms has also attracted much criticism. The are several reasons for this. First, BAME and BME are not specific, and fail to capture the full diversity of people. Second, these acronyms categorise people by skin colour. Third, not many people identify with this term (Milner and Jumbe, 2020). BAME has become a catch-all category that 'conveniently' subsumes identities of people that do not fit into the brackets of 'whiteness' and homogenises diversity.

In response to this challenge, many scholars and activists call for alternative language, for example, 'racially minoritised', as coined by Yasmin Gunaratnam (2012); or 'non-white', 'visible minority', 'Black or Brown', 'Racialised Black or Brown or X', 'Person of Colour', or 'Racialised Person' (Adébísí, 2019). This stance is echoed in the UK Government's response strategy to the Commission on Race and Ethnic Disparities calling for disaggregation and ending use of the term BAME 'to better focus on understanding disparities and outcomes for specific ethnic groups' (HM Government, 2022, p. 31). This recommendation comprises two actions in relation to the term BAME. The first action focuses on ensuring that language can 'communicate more effectively on racial issues' (p. 41) and to stop the aggregation of minoritised ethnicities by government and public sector bodies. The second action relates to improvement of data presentation and interpretation. The strategy recommends referring to 'people from ethnic minority backgrounds', only 'where it is absolutely necessary to draw a binary distinction between the ethnic majority and ethnic minorities' (p. 14).

It is important to acknowledge these debates and criticisms, and where possible to provide or refer to disaggregated data that reflects the richness of cultures and heritages. Unfortunately, many reports and statistics published still aggregate information about racially minoritised people into a single category of BAME, albeit this is slowly changing, or it is at least acknowledged as a limitation. In light of the available literature, rather than reflecting the authors' stance, this book uses the term BAME where this accurately reports on the material referred to, while using specific terms where underlying information allows. In recognition that the term 'ethnic minority' may unhelpfully emphasise the ethnic status of a group, 'minority ethnic' is preferred in order to emphasise the minority status, and is used in the book.

Sex and gender

The UK Equality Act 2010 lists 'sex' as a protected characteristic, and recent focus on equalities has included sex and gender. The Act superseded the Equal Pay Act 1970, which provided men and women equal rights for terms and conditions of employment, including pay. In step with these changes in the legal landscape, much early EDI work at UK universities sought to address pervasive and visible gender or sex inequalities. A notable example is Athena Scientific Women's Academic Network (Athena SWAN, see Chapter 6 in 'equality charters') focusing on efforts to equalise access to opportunities and resources. This aims to lead to parity in outcomes between men and women.

Although not without debate, definitions of sex often refer to biological differences and variety and definitions of gender often refer to social or cultural norms and expectations. The European Institute for Gender Equality defines gender equality as parity between men and women. This definition refers to women's and men's rights, responsibilities and opportunities in relation to their social roles that are independent of sex.

It is worth recognising the long history of scholarship relating to gender, including work to highlight the existence of many, rather than two, genders. For instance, Johanna M. Schmidt (2017) describes how, for almost a century, the concept of a 'third gender' has been explored by anthropologists through examples of cultural or historical contexts where masculinity and femininity transcend Western understandings of gender. Examples include Samoan males, Fa'afafine, who behave in 'feminine' ways (Schmidt, 2016); or the Two Spirit people in North America, who are individuals not necessarily identifying as LGBT, but whose behaviours or beliefs may sometimes be interpreted by others as uncharacteristic of their sex (see Jacobs et al, 1997). Gender is 'performed' through norms, roles and relationships and is communicated to and perceived by the outside world. Philosopher and gender theorist Judith

Butler suggests that there is far more complexity in gender performance beyond the dichotomy of two genders, because there can be other variations of masculinity and femininity. For Butler, gender is also a *'corporeal style*, an "act", as it were, which is both intentional and performative, where "performative" suggests a dramatic and contingent construction of meaning' (1990, p. 190, original author's emphasis). A socially constructed conformity between biological sex, gender, and sexual orientation is a norm and any deviation from this alignment may elicit stigma, discrimination or exclusion. The intersection of discrimination related to gender with discrimination based on deviation from perceived norms presents a challenge that equality initiatives should recognise.

Gender reassignment is also a protected characteristic according to the Equality Act 2010, and it is important to recognise the particular challenges and discrimination that transgender people may experience. For instance, in a qualitative study investigating the common social experiences and minority stressors related to being transgender, Heidi Levitt and Maria Ippolito (2013) emphasised the tensions between transgender people's needs to be respected and valued for who they are and their abilities, and a need to mitigate the dangers of being visible or 'outed' in different contexts. Their work highlighted that the process of negotiating gender in the workplace was complex and stressful for transgender individuals, their competence could be overshadowed by their gender, seeking social support could be risky, and finding safe spaces was elusive.

Sexual orientation

In the UK, LGBTQIA+ is the acronym widely used to refer to members of lesbian, gay, bi, trans, queer, intersex, asexual, and related communities. Other acronyms and terms are understood and used, some with greater breadth. Throughout this book we use LGBTQIA+, unless quoting other authors who use other terminology to reflect a narrower research focus (for example, LGB only).

Sexual orientation and sexuality are multi-layered. Kevin Guyan (2022) explains how sexual identity, sexual attraction and sexual behaviour all are subsumed under the umbrella term of 'sexual orientation', yet they have different meanings. *Sexual identity* describes 'how a person thinks of their sexuality and the identity terms with which they identify' (for example: straight/ heterosexual; gay or lesbian; bisexual; asexual); *sexual attraction* refers to 'sexual and/or romantic feelings' towards one, or more than one, specific sex or gender, or a lack of attraction towards anyone (2022, p. 76). On the other hand, *sexual behaviour* concerns sexual partnering with another sex or gender, with the same sex or gender, or not engaging in sexual behaviour.

However, for the purposes of the legislation, sexual orientation is understood broadly. In data collection for monitoring purposes, questions about sexual orientation capture a respondent's self-perceived sexual identity (Office for National Statistics, 2009). It is worth noting that using sexual identity as a proxy for sexual orientation is problematic, since studies have shown that the reported population size of minority sexual orientations varies depending on which dimension is used in surveys. Put simply, what people say they feel or what they say they do does not necessarily match what they actually do: identity, attraction and behaviour are different aspects of sexuality (Guyan, 2022). The OECD (2019) reported that measuring sexual orientation through sexual behaviour instead of sexual identity increases the reported LGB population by 70 per cent. Measurements that use sexual attraction as a criterion to define sexual orientation instead of identity result in twice as many LGB respondents. However, as sexual identity is believed to be most closely related to experiences of disadvantage and discrimination, sexual identity is most commonly used to assess difference.

Experiences of discrimination of LGBTQIA+ individuals and communities can be understood through the concept of 'heteronormativity', whereby 'institutions, structures of understanding and practical orientations [...] make heterosexuality' the dominant and privileged 'organised' orientation (Berlant and Warner, 1998, p. 548). Any deviation from heteronormativity can result in social marginalisation and oppression, and still in more than 70 countries same-sex relationships are criminalised (Stonewall, 2022).

More action is needed in relation to injustices affecting individual careers, equal pay, and advancement of LGBTQIA+ individuals (Drydakis, 2015). Early, ground-breaking work in this context—research on LGBQ-identifying workers' experiences of paid employment through the 1980s and 1990s—suggested key lessons important for equality work today. First, studies have shown that these individuals constitute a sizeable portion (visible or not) of the workforce. Second, non-heterosexual employees are disadvantaged in relation to career outcomes as well as their personal, material and psychological spheres. Third, the cascading negative effects of individually experienced inequalities impact on employer's organisational performance (Ozturk, 2011). The power of exclusionary heteronormativity is demonstrated in studies that focus on the impact of disclosure of sexual orientation. For example, in academic contexts, individuals who are both perceived as and identify as LGBTQ are faced with loss of self-determination, privacy and safety; whereas individuals who identify as LGBTQ but are not 'out' are most negatively impacted as they have to also navigate disclosures (Beagan et al, 2021). These experiences have negative cumulative effects on individual experiences of career, inclusion and belonging.

Additional categories of diversity

EDI-related work within research and higher education (HE) should be guided by mutual understanding, dialogue and participation. As well as presenting the five aspects of diversity and current debates, it is suggested that EDI advocates and practitioners practise reflection when considering interventions to address inequalities and disadvantage. As such, in Table 1.1 we offer the following questions as a starting point to a meaningful conversation and deliberation of EDI efforts.

To provide a more comprehensive picture of diversity categories that guide understanding of issues of equality and inclusion, a broader diversity framework is briefly presented here, partially inspired by the PROGRESS-Plus framework (O'Neill et al, 2014). This framework is adapted by the authors to research careers by drawing on topical debates. This is to encourage the reader to consider how additional layers of social difference may determine equality and equity in research careers and how these layers may intersect and combine to amplify bias and disadvantage.

Table 1.2 captures these additional dimensions as well as those discussed earlier and suggests reflexive questions related to research careers. As well as reflecting about the questions in the table, it is important that members of the research ecosystem are alert to the potential for the emergence of new categories of diversity, difference and adversity that are not yet captured here. These could be related to new political and economic developments, such as future impacts on individuals and research from institutional reforms, immigration, war, international health emergencies and geopolitical shifts.

Intersectionality: multiple categories of diversity

The term 'intersectionality' is generally thought to have originated from Kimberlé Crenshaw's critical analysis of anti-discrimination law in the US in the late 1980s. Crenshaw, who primarily focused on the context of law, described how the discrimination that a person might experience is exacerbated when they identify with more than one

Table 1.1: Reflexive questions to guide EDI interventions

Disability	• Have researchers had access to and appropriate support for disabilities on an individualised basis as well as more broadly without the need to declare disability, for example, designing inclusive environments as a default? • Have barriers or inadequate adjustments been considered in terms of long-term impact on a researchers' careers? • Has there been a meaningful conversation with disabled researchers: 'nothing about us without us'?
Neurodiversity	• Are research-related activities designed in ways that mean they are appropriate for neurodiverse researchers, for instance: conference attendance, presentations, talks, networking, writing, interviews? • Has preferred language been discussed, considered and respected, for instance: some prefer the term 'autistic person', others 'person with autism'?
Race and ethnicity	• How does the institutional, departmental or group research culture support minority ethnic researchers to ensure inclusion, for instance in collaborations, networks and spaces? • What are the potential impacts of inadequate inclusion into a research environment on minority ethnic researchers and their careers? • How can minority ethnic researchers be supported to lead research funding applications, large grants, and bids to increase success rates to address underrepresentation?
Sex and gender	• What specificities and inequalities exist in particular fields or disciplines—are these gendered, or horizontally or vertically segregated, and how do they shape researchers' careers? • How can the impact of gender biases and stereotypes be minimised in selection, promotion, funding decision processes, peer review, authorship and performance evaluation?
Sexual orientation	• How does the social context of a LGBTQIA+ researcher support or create barriers to wellbeing, research performance, networking and collaborations (both internal and external)? • What specificities, stereotypes and exclusions exist in particular fields or disciplines—hypermasculinity, heteronormativity, cisnormativity? • Is the research environment conducive to an open approach to sexual orientation?

Table 1.2: Reflexive questions encompassing additional categories of diversity to guide EDI interventions

Educational and institutional affiliation	• How does the professional context in which researchers work impact on career pathways, for instance, an organisation's esteem, strategy, resources and infrastructure? • Have researchers had appropriate support from their research organisations to avail of mentoring, development, training, opportunities and networking?
Professional and discipline affiliation	• How might the popularity or public appeal of certain research topics impact on careers? • How is disciplinary or interdisciplinary impacting on career trajectories? • Are disciplines 'gendered' or patterned in other ways and how may this be impacting on careers?
Socio-economic status and social capital	• Have researchers not had access to social and professional networks to help them access certain privileges or accelerate their career? • How might researchers' social class or economic backgrounds be impacting on whether an organisation and research community includes them?
Culture, language, religion, and country of origin	• To what extent might researchers' culture, language, or religion impact on whether an organisation and research community includes them, for instance for collaboration, collegiality and networking? • How might immigration status be influencing researchers' career decisions, productivity and trajectory?
Occupational classification	• Have fixed-term research contracts impacted on researchers' outputs, career and stability (predictability), and how? • How have researchers' teaching loads, clinical or professional practice, industry engagement, and administrative duties affected research productivity?
Parental/paternal carer status	• For researchers who have young children, family with disability or care needs, how is research output and aspirations impacted and what can be done to support them?

(continued)

Table 1.2: Reflexive questions encompassing additional categories of diversity to guide EDI interventions (continued)

Education	• Where did researchers gain education? Consider (inter)nationality, culture, prestige of educational settings as well as the 'home' discipline.
Age	• How is the traditional notion of 'research time' impacting how competence of researchers is viewed? • What stereotypes and assumptions about age might be impacting on researchers' access to opportunities? • How might age in intersection with sex pose challenges to research careers, for example, menopause?
Marriage and civil partnership	• How might marriage, civil partnership or singlehood affect researchers' ability to be geographically mobile, secure international grants or to remain in a chosen place/institution?
Pregnancy and maternity	• How might pregnancy and related considerations be impacting on researchers, for instance breastfeeding? • Have researchers about to take on, or returning from, maternity leave been given appropriate support for their research? • How have periods of maternity leave impacted researchers' outputs and how is this taken into account in selection, promotion, performance evaluation and funding decisions?
Any special circumstances?	• How have issues such as illness, family circumstances, family tragedy, effects of the COVID-19 pandemic, major life events, been accounted for in selection, promotion, performance evaluation and funding decisions?

category of social difference. In the original court cases that Crenshaw analysed, this was in relation to race intersecting with gender. Specifically, African American women claimed that their employment experiences of discrimination were the direct result of being both Black and female, not just being Black or female (Crenshaw, 1989). Although the spirit of intersectionality predated Crenshaw's work, including in Black feminism, legal frameworks at the time were unable to grasp this interconnectedness[2] (Rodó-Zárate, 2020). As such, Crenshaw argued that inequalities, discrimination, and the benefits or privileges that individuals experience depend on several different characteristics and cannot be explained by one identity, category or location.

Nira Yuval-Davis (2015) also introduced 'situated intersectionality' to attend to the role of context and particular social and historical configurations to examine the complexity of social inequalities. Intersectionality therefore provides a helpful lens to reflect and understand that inequality can be the result of other various compounding social factors simultaneously: gender, race, ethnicity, age, disability, sexual orientation, religion or belief, gender reassignment and socio-economic status. The concept of intersectionality has contributed to areas beyond Crenshaw's legal context, such as research, scholarship, education, health, activism and social justice practice. For example, practitioners may need to address disadvantage and related problems in their work, and Patricia Hill Collins argues (2015) that educators, social workers, policy advocates, university support staff, community organisers and researchers are often exposed to complex social inequalities. Being at the forefront of potential experiences of inequality conditions, these practitioners are able to recognise and then reflexively respond to intersectionality. For practitioners within research and academic settings, intersectionality can help them to understand how social justice work can be effectively performed through carefully considered practices, constant (re-)evaluation, assessment, and reflection about interventions that are most likely to lead to positive social change (Collins, 2015).

Explorations of lived experiences of intersectionality show that the intersection of age, gender, ethnicity and religion amplifies the separate social dimensions. For instance, Rashida Bibi's study explored mature British Muslim South Asian women students who had to negotiate gendered, cultural and religious realms to access academic careers (Oman et al, 2015). For instance, one participant described how she perceived other people's evaluations of her: 'maybe it's just me ... but I remember walking home and people thinking she should be carrying babies not books' (Bibi, 2015, cited in Oman et al, 2015). Bibi's research indicates how multiple identities need to be managed to satisfy expectations of women's community, family and academic institutions as well as stereotypes of Muslim women.

Intersectionality is not reflected in legislation in the UK or the European Union. Even though section 14 of the UK Equality Act 2010 had been drafted to provide a possibility of combined direct discrimination based on dual characteristics, in 2022 this section was still prospective. The UK government has rejected the possibility to combine more than three characteristics as this would prove 'unduly complex and burdensome for employers and service providers' (Government Equalities Office, 2009, p. 43). This means that intersectional claims are instead dealt with on the basis of available case law (Bourne, 2020).

Diversity data collection

Data collection is the first step to analysing staff and student diversity within research and academic organisations before any action plans can be designed to address underrepresentation of specific groups. In the UK, this occurs through submission of diversity monitoring data to the Higher Education Statistics Agency (HESA) that collects and disseminates the data that forms part of the public sector compliance enabling academic and research institutions to carry out their public functions. Some of the organisations that make use of the data include the UK Government, UK Research & Innovation, and Office for Students, among others. Diversity data collection differs internationally, and collection of information in relation to race and ethnicity information, but also other characteristics, may be very complex in certain countries (European Commission, 2021a). UKRI reviews of EDI in research and innovation offer valuable information about enhancing data and disclosure in the UK (Guyan and Douglas Oloyede, 2019) and international contexts (Moody and Aldercotte, 2019), and the guidance presented here focuses on the UK context.

HESA requires HEIs to submit diversity data in a prescribed format; however, there is a level of flexibility to allow organisations to tailor data collection by narrowing the foci of data to reflect their individual contexts. Advance HE (2022) offers guidance for diversity data collection that reflects the current legal position, and it complies with the Equality and Human Rights Commission guidance. However, Advance HE recognises that diversity data collection is 'complex and contentious', thus research and academic organisations may wish to adapt this guidance to reflect their own individual needs and requirements (Advance HE, 2022). Data collection is not an apolitical and unbiased practice. As Kevin Guyan argues, data collection shines light on certain identity characteristics, but not others, which is the result of decisions about who 'should be counted'. This can lead to some individuals remaining unaccounted for and not legitimised (Guyan, 2022).

Further helpful recommendations about diversity data collection are provided by the Diversity and Inclusion Survey question guidance (DAISY) (Molyneux and Hunt, 2022). DAISY is complemented by guidance for improvement of response rates (Hunt, 2021) and was developed in collaboration between Equality, Diversity and Inclusion in Science and Health Coalition (EDIS) and the Wellcome Trust (WT), a

UK-based charity that operates globally as a major funder of health research. DAISY is an ongoing project that consults key stakeholders and expert bodies, with the overall aim of championing principles of inclusive practice in diversity data collection. For example, the approach includes embedding the principles of anti-ableism, recognising the social model of disability, additional socio-economic status questions, and more inclusive sexual orientation questions.

Summary

- 'Equality, diversity and inclusion' and its language, focus, and meaning are in a constant state of flux. This change takes place in the light of the development in empirical evidence and knowledge, and the general intellectual, moral, political and cultural climate in society and the research ecosystem.
- EDI practitioners, researchers and stakeholders need to stay up to date with recent developments and critical debates.
- Anyone working in EDI should consider how best to engage in constant dialogue with the communities affected by EDI initiatives.
- The nine protected characteristics are a starting point from which to develop awareness of multiple additional dimensions of diversity, difference and adversity.
- Application of an intersectional perspective provides a powerful way to consider how multiple categories of diversity amplify disadvantage.
- Data collection for diversity monitoring purposes is a starting point for equality interventions and action plans. Organisations that wish to collect diversity data beyond the minimum statutory requirement are encouraged to do so in keeping with inclusive data collection guidelines, such as DAISY.

TWO

The current context of research careers in health and biomedicine

This chapter presents the current shape of research careers, including the context of health and biomedical sciences and the scale and influence of the area as an employer. The challenges and opportunities for institutional change are presented as well as the landscape of research careers that already imposes obstacles for progression within the research community. Some of the available literature focuses more broadly on careers in science, technology, engineering, mathematics (STEM) and medicine (STEMM), but nonetheless provides an important and relevant data framework for careers in health and biomedical research. The chapter prepares the ground for guidance about how to address key EDI concepts and issues in later chapters.

The shape of research careers

This chapter explores research careers in health and biomedicine. Although the focus is on researchers, the work of researchers is only possible in light of myriad other, equally important, roles in the research system, including officers and administrators who work in research impact and development, governance, funding, public engagement, enterprise, commercialisation and investment, laboratory technicians, human resources staff, to name a few. Professional support staff members do not only 'support' the research and academic system, but continuously construct, maintain and improve practices and processes that are vital to the whole system. Support staff also add to researchers' careers, professional development and wellbeing, starting in the earliest stages of student life and into the many levels of research careers.

Increasingly, EDI in research and academic institutions in the UK is incorporated in formalised roles within human resources (HR) departments as well as at departmental, school or faculty level through roles such as EDI champions or leads. Individually and collectively, such stakeholders have the power to make a difference and help to shape research careers.

What are research careers?

The Organisation for Economic Co-operation and Development (OECD) defines 'scientific researchers' as people who are professionally trained to undertake 'any creative systematic activity [...] in order to increase the stock of knowledge, including knowledge of humankind, culture and society, and to devise new applications of available knowledge' (OECD, 2015, p. 44). In this conceptualisation, science is broad, encompassing all disciplines and fields rather than solely natural or allied science. 'Research career' is defined as a trajectory of research-based employment that can take place within and across academia, research councils and research institutes, industry, government laboratories and agencies. People working in health and biomedical research follow a variety of different career pathways. Careers develop and change and can be impacted by individual, social and economic circumstances. The word 'career' comes from the Latin 'carraria', a carriageway or road leading somewhere, indicating a sense of ongoing progression and development. Although conventional academic research careers can be seen as based on linear progression and promotion, we know that the reality for many is different, and in many other sectors careers are frequently understood as decidedly nonlinear.

There is no uniform metric to assess what career 'success' looks like in health and biomedical research, as success will always depend on an individual's own aspirations, including their plans for living their life outside their research work. Also, a pathway to success may look different according to discipline, field of research, and research environment. Broadly speaking, members of the research community are expected to design and deliver research that helps to move their field forward, in adherence to professional and regulatory standards, using their skills and knowledge to do so. The types of outputs that health and biomedical researchers might produce along the way are likely to include some or all of the following: conference presentations, published research articles and research reports, pre-registered protocols, shared datasets and products such as code, books and patents. However, appropriateness of outputs varies greatly by field and discipline; while a book may be an outstanding product in one field, in another a bank of openly available data or code might be highly prized. For all disciplines, outputs take place in the context of much broader work and contribution. Initiatives to recognise the broad range of contributions that researchers make is addressed in Chapter 6, where interventions including the narrative CV are described.

Enabling people to reach their potential as capable researchers who thrive professionally is crucial to the success of research organisations. The performance of organisations is shaped by careers of its employees (Higgins and Dillon, 2007). For instance, individuals who 'traverse' various employing organisations bring in their human and social capital, such as work and educational experience, connections, networks and interorganisational affiliations. Such resources are seen as 'assets' that can benefit the organisation (Higgins and Dillon, 2007).

These tacit commodities as well as the researchers' specialist knowledge and acumen are leveraged by members of the research community to deliver the research that is assessed. For instance, UK universities work within a system of performance-based funding, such as the Research Excellence Framework and grant-awarding processes. Internationally, similar approaches to research assessment are in place, and are refined and develop over time (Sivertsen, 2017). In the marketised higher education landscape, these approaches can affect institutions' attractiveness to students and research collaborations, which in turn may enhance or protect universities' financial security (McGettigan, 2013).

A focus on excellence in relation to diversity is an international phenomenon. In a review of certification and awarding systems for gender equality and related inclusivity schemes across the EU's 27 countries and Australia, Iceland, Norway, UK, Switzerland and the US, Giulia Nason and Maria Sangiuliano (2020) showed that 27 per cent of these schemes were explicitly linked to excellence either in management, education or research and framed as part of the strategy. The connection between gender equality, inclusion and excellence was more often explicitly made in the context of management, also including business and public administration certification and award systems. These schemes were predominantly concerned with the inclusion of women in leadership positions and mechanisms for supporting women's careers, linking equality with excellence in management practices. The case for diversity may also be linked to movements that achieve and assess research quality or excellence, as described in the next chapter. Importantly, the quality of careers also has individual-level consequences, such as financial (in)security, professional fulfilment, personal identity, health and wellbeing (Lieff, 2009; Steinþórsdóttir et al, 2019; Cardel et al, 2020; Hollywood et al, 2020).

International mobility continues to be an essential part of knowledge exchange and related progress in research.

Globalisation and massification of research mean that researchers' careers unfold internationally or at least nationally. Mobility across nations to conduct research and collaborate significantly increases one's international networks, advances research skills, boosts career and recognition in the research community (UKRI, 2022c). However, international mobility is not realistic for everyone. Nearly 80 per cent of post-PhD individuals who decided not to move internationally did so for personal and family reasons, and this has risen from 67 per cent over seven years (European Commission, 2021c). Moreover, international mobility can cause complications relating to the recognition of qualifications, portability of pension, access to state benefits, language skills or immigration rules (OECD, 2021). On a relational level, national and international mobility moves people away from their local communities and support networks, and can complicate work–life dimensions.

Research careers are also shaped by the relatively recent oversupply of research students (OECD, 2021), with the most rapid rise in PhD candidates seen in US life sciences (Gould, 2015). For instance, in the US, 30–40 per cent of life science PhD recipients graduated without a secure job or postdoctoral commitment between 2009 and 2019 (National Center for Science and Engineering Statistics (NCSES, 2020). In response to this trend, scholars have called for internal support mechanisms to ensure that PhD scientists are well-equipped to pursue alternative career paths (Lee et al 2010; Zimmerman, 2018). Universities are producing more postgraduate researchers, a situation of supply exceeding demand which is further exacerbated by mechanisms for research funding. Although, in the US and UK most research is conducted in universities, the OECD reports that tenured academic roles that combine pedagogy and research are now rare in some systems, due to replacement of long-term funding with core resource allocation by short-term project funding. This change feeds precarious employment arrangements and encourages academic and research institutions 'to rest their research prowess on a few tenured star researchers' who lead research groups of doctoral and postdoctoral researchers running the operation of projects (OECD, 2021, p. 16). Instead of permanent academic roles for individuals after their PhDs, the system operates through temporary contracts that align with the length of the grant period. This pragmatic agility benefits research organisations, while postdoctoral researchers are unable to enjoy long-term security (Woolston, 2021). There is a significant chasm between higher education (HE) and non-HE contracts when comparing destinations of leavers from HE: just under 15 per cent of those working in non-HE research were employed on short-term contracts, while for individuals working in HE research this figure was significant, at 75 per cent (CRAC, 2019). Such precarity has disproportionally impacted women researchers leaving HE research roles, as fixed-term contracts do not provide job security and work–life balance. Interestingly, among those researchers who left HE, it was the biological and biomedical researchers who found it most challenging to adjust to a new culture and the lack of flexibility in working hours (Haynes et al, 2016).

This kind of employers' market creates an obvious disadvantage at the individual level and provides abundance for employing institutions recruiting and selecting diverse talent. However, as we interrogate diversity statistics it becomes clear that this is not the case. Historically, scientific activity was largely carried out by those with social and socio-economic privilege, particularly by those who were White and male (Schiebinger, 1987; Dias Lopes and Wakeling, 2022). It is argued that one of the reasons that inequalities persist is because individuals are embedded in institutions that were often built and developed on inequality regimes (Acker, 2006), racism and patriarchy (Gabriel and Tate, 2017), and other forms of oppression. The inroads that White women, and people from disadvantaged socio-economic backgrounds, have made into jobs in HE are relatively recent. For some, particularly from Black (and Black British Caribbean and African) and Asian (Pakistani and Bangladeshi) backgrounds in the UK, rates of entry into academic and research jobs have been low and slow (HESA, 2022c). Speaking of Black and Brown academic women, Deborah Gabriel states that '[y]ou don't need to review the latest figures from the Higher Education Statistics Agency to know that we are few in number. Just visit any university' (Gabriel, 2017, p. 1).

Legacies of structures, systems, policies and practices in institutions all require continuous and gradual change through targeted interventions, equality plan implementation, equality impact assessment and culture change. Academic and research careers are also situated within broader networks of stakeholders, including research teams, groups, peer networks, communities of practice, funders, employers, research communities and national cultures. Social inequalities may be produced and reproduced in all these contexts.

Research careers in health and biomedical sciences

The COVID-19 pandemic, and the ensuing rapid vaccination programmes across the globe, placed a sharp focus on the importance of health and biomedical science and careers. Scientists, academics and technicians have been working tirelessly to produce and deliver safe and effective vaccines. The Institute of Biomedical Science highlights that, even though the biomedical scientists only constitute around 5 per cent of the UK workforce, they are crucial to the UK healthcare industry. For example, around 70 per cent of all diagnoses are attributed to the work of biomedical scientists (Institute of Biomedical Science, 2022). John-Arne Røttingen and colleagues (2013) report that total global investments in both private and public sector health research and development (R&D) reached US$240 billion in 2009. It should be noted that most of the investment (US$214 billion) was in high-income countries. Specifically, most health R&D investments came from the business sector (60 per cent) and half of that from the public sector (Røttingen et al, 2013).

Despite the scale of investment in health and biomedicine as a sector, career paths are highly competitive and require perseverance. According to the not-for-profit venture '80,000 Hours'—that advises people on how to choose a career that provides a high societal impact—careers in biomedical research are likely to produce large returns to society, but individuals wishing to continue into academic may face many challenges (Duda, 2015). These careers require resilience and luck: it takes long periods of time to qualify, securing a permanent post is highly competitive, and relatively few people achieve tenured or permanent professor roles. It is depressing but hardly surprising that researchers are advised to boost their resilience to protect themselves from 'developing clinical levels of anxiety and depression' (Gloria and Steinhardt, 2016, p. 155). However, researchers' resilience is also deeply shaped by the research ecosystem and its stakeholders. This occurs through embedded activities such as peer review influencing researchers' decisions whether to persist or switch away from academia (Derrick et al, 2022). This finding strengthens the argument against deficit model interventions and expectations put on researchers to 'toughen up' to survive, and instead calls for holistic approaches to nurturing research careers and making the research culture 'kinder' (Derrick, 2020). Moreover, although science careers can be lucrative for established researchers, the early stages of careers can be particularly challenging when employment involves short, fixed-term contracts. For instance, relatively low-paid postgraduate researchers are exposed to the impact of broader contexts, such as cost of living crises in light of international economic trends (Woolston, 2022).

An insights and analysis report based on data from the UK HESA Graduate Outcome survey shows relatively low persistence rates. It suggests that only 20 per cent of biology graduates follow a career path into a scientific occupation, while over 30 per cent end up in non-graduate jobs (AGCAS, 2022). Only 3.5 per cent of people with a science PhD succeed in securing a permanent research position in academia, with only 0.45 per cent progressing to professorship, and 17 per cent continuing their professional careers in non-university-based research roles (Royal Society, 2010).

The career intentions of doctoral researchers have been explored in detail in a report by Robin Mellors-Bourne and colleagues (2012). The report is useful, not least because of its scale: over 4,000 respondents based in the UK took part. Respondents were aligned with a range of disciplines and fields, including 686 from biomedical and 620 from biological sciences. The proportion of doctoral researchers seeking research roles outside of HE was much higher in biological (55 per cent) and biomedical (34 per cent) sciences than other discipline groups. For comparison, only 6, 3 and 1 per cent of respondents in social sciences, arts and humanities, and education, respectively, intended to work outside HE. Additionally, over 30 per cent of biomedical sciences researchers sought other common doctoral occupations. Specifically, 85 per cent of the 'biomedical sciences respondents expected to work in the health sector and 88 per cent in a health and social care job function' (2012, p. 8). Decisions not to pursue an occupation in research in biomedicine were mostly down to perceptions that 'there were too few career opportunities in this field' and that doing other work was more lucrative (Mellors-Bourne et al, 2012, p. 13). In a more recent report by CRAC (2022), it was shown that biomedical doctoral researchers were much more likely to work in doctoral occupations other than HE or non-HE research or teaching, such as in health roles. The percentage was as high as 43.6 per cent for biomedical sciences, compared to just under 15 per cent for biological science (CRAC, 2022).

This shows vast differentiation and career expectations between the disciplines and confirms how uncertain biomedicine researchers feel about pursuing a career in academia.

Similar to other STEM fields, currently there are more individuals in biomedical PhD training and graduating than available academic faculty and research positions. This means that biomedical PhD holders need to be open minded, think 'outside of academia', and be flexible in how they plan their future career paths (Zimmerman, 2018). Indeed, Hsing-fen Lee and colleagues (2010) suggest that scientists with PhDs are more likely to work in industry, while academia has become a secondary employment sector. Industry becoming the major employer can be observed both in the UK and in the US. For this to be seamlessly accomplished, studies show that the hosting faculties need to help PhD trainees—for example, connecting them to opportunity-bringing networks, redefining what it means to be successful, destigmatising alternative paths, and helping to define academic and career goals early in the training process.

Robin Mellors-Bourne and colleagues (2012) highlighted a moderate difference by gender in relation to career advice. More women than men reported that 'they would have benefited from additional careers advice and support' (69 per cent compared to 59 per cent respectively) (2012, p. 28). This suggests that some targeted interventions might be needed (Mellors-Bourne et al, 2012). Funding bodies also have a role to play in acknowledging and promoting scientists in non-academic sectors (Zimmerman, 2018). This includes valuing non-conventional paths and accomplishments, which is increasingly recognised through interventions such as narrative CV (for example, the Royal Society's Resumé for Researchers; see Chapter 6). This is sorely needed: for instance, as many as 42 per cent of the CRAC report respondents entered biomedical doctoral research studies from other permanent, casual or temporary jobs and other settings, rather than immediately after their under- or postgraduate education. The picture was more overwhelming for those studying part-time, of whom more than 82 per cent entered biomedical doctoral research from employment or other settings (Mellors-Bourne et al, 2012).

EDI in research careers

Diversity among researchers should be the starting point in conversations about EDI in research careers, and both the local and wider context need to be acknowledged. For example, to what extent does the diversity in an institution reflect the diversity of a local community, or the population at the national level, or the groups that research aims to serve? How diverse are the students, academics, lab technicians and support staff? Diversity among these groups is typically explored when embarking on an equality plan implementation to enable self-assessment and comparison with other institutions. For example, EIGE recommends that the most effective process starts with analysis and assessment of the current context in the institution, and then developing a well-informed action plan that attends to its specificities (EIGE, 2016a). This is normally achieved by comparing given institution to national benchmarks calculated from the averages of other institutions, before attempting to design targeted action plans if these plans are to be meaningful.

This is because we should not assume that all institutions are experiencing the same problems. For instance, University College London (UCL) illustrates great variation between institutions. In its 2020 race equality charter application, UCL reported that its location in London meant it was more ethnically diverse than the national HEIs and the Russell Group average. However, despite its efforts, UCL was around 2 per cent less diverse than other institutions within London. The reasons given by the UCL self-assessment team for this were varied, such as different institutional contexts, dominant disciplines, and recruitment from different talent pools (UCL, 2020).

Recent evidence on diversity of individuals in health and biomedical research careers is presented later in this chapter. However, generally in many countries both in research and academia across STEMM, there is persistent underrepresentation of women, minority ethnic, disabled people (including people with long-term health conditions), and individuals from underprivileged socio-economic backgrounds. Indeed, data from HE and research funders indicate that gender, race and ethnicity, but also institutional affiliation and socio-economic background, shape researchers' career paths and their ability to achieve what is traditionally perceived as 'academic' or 'research' career success (Williams et al, 2019; Dias Lopes and Wakeling, 2022).

In the UK, the Government recognises this and engages in various initiatives to tackle this challenge. For example, in 2018, the All-Party Parliamentary Group (APPG)[1] on Diversity and Inclusion in Science, Technology, Engineering and Maths (STEM) was established to 'promote the inclusion and progression of marginalised people in STEM' by nurturing diversity through policy. It recognised that STEM as a sector is not representative of the UK population in general and the way to change this was to consult with as many stakeholders as possible in tackling underrepresentation collaboratively.

In the Equity in the STEM workforce report published in 2021, Chair of APPG on Diversity and Inclusion in STEM, Chi Onwurah MP, introduces the report with the words: 'being different in any profession or job is tiring, you face people's stereotypes rather than being judged on your actual experiences and ability' (British Science Association, 2021). Reflecting on her own engineering career in the 1980s and 1990s, and her intersectional social categories as a Black woman from a working-class background from the north of England, she recognised that, four decades later, many minoritised communities still do not feel that they belong in the STEM workforce. The UK Government also acknowledged these challenges in its R&D People and Culture Strategy, and recognised that if research culture 'is not seen as open and inclusive' this 'puts off and drives out talent, preventing individuals from producing their very best work' (BEIS, 2021, p. 11).

There is much at stake for the UK, since the government plans are ambitious for R&D. The demand for scientists in the R&D sector has been growing. This demand was stated in the White Paper 'Innovation Nation' (Department for Innovation,

Universities & Skills, 2008); and recently reaffirmed in the R&D People and Culture Strategy (BEIS, 2021). The target for research and development intensity in the UK is to increase from 1.7 to 2.4 per cent by 2030 (BEIS, 2021, p. 5). It is estimated that the R&D sector will require at least an extra 150,000 researchers and technicians across public, industrial and all other sectors by 2030 to sustain this ambition. To achieve this, research careers need to become attractive, inclusive, broader, and made less precarious, especially in the early stages of career development (BEIS, 2021).

The 'leaky pipeline' and diversity

The metaphor of the 'leaky pipeline' has been used in many parts of STEM to describe the progressive loss of individuals to science. The idea was originally developed to convey how women drop out from careers instead of progressing along the pipeline from student to leadership. The leaky pipeline is now used to describe loss of members of many groups, including minority ethic individuals, along the career pathway. The pipeline also indicates that others do stay in the pipeline; in total this leads to underrepresentation of some groups at leadership levels.

The leaky pipeline may be convenient for illustration, but it is controversial. First, it obscures the root causes of attrition of women and minority ethnic individuals from the research ecosystem. It is not clear why the loss of talent takes place, whether leakage happens because of structures, systems, gatekeepers, support mechanisms, or individuals themselves. Conversely, as the pipeline contains a filter and leaks selectively, it provides benefit to certain individuals. Second, the pipeline model tends to separate research in academic settings from settings that are research performing, including private and third sector entities. When individuals leave academia and move into another type of organisation, they are not necessarily lost to science; instead they may be contributing equally, or more, in another setting. This is important, as placing value on such work helps to frame research as a broad ecosystem, and acknowledges that much science takes place outside academia and its pipeline. International efforts to incorporate this perspective are exemplified by new initiatives, such as the narrative CV, which recognises a broader range of career paths and contributions (see Chapter 6).

To understand the complexities of leakage from the pipeline, Petchey and colleagues draw attention to the 'headwinds' and 'tailwinds' of implicit (unconscious) bias that affect people differently. Bias in this sense can provide advantage to some individuals while delivering disadvantage to others (Petchey et al, nd). Petchey and colleagues' work was conducted at the Faculty of Science at the University of Zurich, Switzerland, and has global relevance with a step forward in how the 'leaky pipeline' is understood and addressed. Importantly, understanding the presence of leaks through the metaphor of headwinds and tailwinds encourages focus on individuals' careers in the contexts of the ecosystem in which they work. This can include looking at the role of broader discrimination, stereotyping, availability of resources and other factors that impact on opportunities. These opportunities—or lack thereof—can take place in lecture rooms, labs, job recruitment panel, grant-awarding committees, in collaborations, in networks, in article authorship decisions, and in places where career-enhancing or inhibiting practices occur. Cumulative advantage or disadvantage can enhance or slow the pace of researchers' career progression and ultimately influence who stays and thrives in research.

Diversity data

In the broader context of STEM, HE data shows that members of the Black minority ethnic group are at particular risk of attrition. Even though Black UK-domiciled students are well represented in undergraduate science courses, they have the highest non-completion rates among all minority ethnic populations. They are more likely to leave both first degree and postgraduate STEM courses with no award. In contrast, Asian, mixed and White female students are most likely to complete their first and postgraduate degrees and achieve first or upper second-class honours, thereby effectively lining the STEM 'pipeline'. And yet, clear vertical segregation occurs for both female and minority ethnic staff. The percentage for senior management, heads of school, senior function heads, and professors is highest for White men (19.5 per cent) and lowest for women (5.4 per cent), and especially low for Black (2.6 per cent) and Asian (4.2 per cent) women (Joice and Tetlow, 2020).

There is a clear and urgent need to equalise Black scientists' access to the research ecosystem, which is evidenced by data illustrating how the representation of Black individuals decreases at each level in academia. Comparing the numbers of Black students with available census data of the wider UK population shows promise at undergraduate levels, where the ratio of minority ethnic groups is higher than in the general population. However, this ratio drops for postgraduate students, and for academic staff and professors it decreases well below the general UK population rates (Gibney, 2022). It is worth noting that, unlike student populations, academic and research staff data collection does include individuals coming from outside the UK; therefore, these figures might be even lower for UK-origin staff. Broad ethnicity groupings in diversity analyses hide nuanced underrepresentation, especially among academic staff of Pakistani, Bangladeshi, Black Caribbean, and Black 'Other' ethnicities (Gibney, 2022).

Narrowing the focus to biomedical and health sciences, the total numbers of all postgraduate students in the UK qualifying stood at 1,355 and 2,135 respectively in 2020–21 (HESA, 2022f). In terms of gender, women broadly outnumbered men in all postgraduate levels, including doctorates across medicine, and subjects allied to medicine, dentistry and biological and sport sciences (HESA, 2022a). Limiting these data further by ethnicity brings into sharp focus the homogeneity of the cohorts: 66 per cent of biomedicine and 59 per cent health science students were White in 2020–21. When these data are further disaggregated into Black and Asian ethnicities, the figures are stark: Asian students would only just fill a typically sized cinema auditorium, at 205 and 295 for biomedicine and health respectively; and Black students would barely fill half, at 70 and 145 respectively (HESA, 2022e). Percentages were particularly low for Black doctoral graduates compared to different ethnicities in biomedical sciences—only 2 per cent of Black, compared with 3.4 per cent mixed, 11 per cent Asian, and 81.2 per cent White graduates (CRAC, 2022).

These data show that, although there is an oversupply of qualified people and thus a large pool of potential candidates to pursue health and biomedical careers, this oversupply is homogenous. Moreover, this homogeneity is particularly visible at the higher seniority levels within institutions, as it is in the case of broader STEM. For instance, women academic staff outnumbered men in clinical medicine and other health sciences, apart from clinical dentistry, where men only just outnumbered women (HESA, 2022b). However, gender-disaggregated statistics for staff level provided by HESA (2021) for UK academic staff in Medicine, Dentistry and Health provide evidence for both horizontal and vertical segregation. That is, women are much less likely to occupy the highest paying professorial positions, and tend to be located across lower contract levels, and spread in teaching-only positions.

Research activities and funding are cornerstones of research career. Data relating to funding outcomes is a particularly rich source of information relating to diversity. The fluctuating success of women scientists in achievement of research funding has been demonstrated by Mulvey and colleagues (2022) who looked at 4,388 applications to the National Institute for Health and Care Research (NIHR) Academy, which develops and coordinates NIHR academic training, career development and research capacity. Mulvey and colleagues wanted to establish what was behind application numbers and success rates by looking at the interrelationships of applicant characteristics. Interestingly, the number of female applicants was lower in the more senior award schemes (such as research professorships), but it was equal with male applicants in predoctoral awards, suggesting a trend of vertical segregation. Men were almost twice as likely to apply for the research professor awards but, even though the number of successful applications from men was higher than from women at senior award levels, the authors computed this was because more men submitted applications for these awards (Mulvey et al, 2022). The authors suggest that, if more women were at the right seniority level and able to apply for these awards, they would be as successful as their male counterparts. As women are not, consequently, reaching the highest staff levels, there is a gender gap in senior award levels.

Diversity data from UKRI between 2015–16 and 2019–20 showed that the proportion of awardees from minority ethnic groups had increased over the previous six years. Nonetheless, representation of minority ethnic groups was relatively low at all levels, particularly among principal investigators (12 per cent), compared with co-investigators (18 per cent) and fellows (17 per cent) (UKRI, 2021a). Variations in funding success rates vary across different ethnicities. For example, the UKRI detailed ethnicity analysis from different research councils for the period between 2015–16 and 2019–20 showed that some minority ethnic groups were more successful than others, reinforcing the argument against aggregating ethnicity into BAME groupings. For example, the share of principal investigator awardees from the Black and Bangladeshi ethnicities was lower than their labour market and HESA academic staff shares, securing fewer grants than their Asian and White counterparts. On the other hand, the share of principal investigator awardees and researchers from the Chinese ethnicity was higher than their labour market share (UKRI, 2021a). Therefore, it is crucial to recognise different outcomes for specific minority ethnic groups. Reasons for this 'seniority drop', and an 'unbalanced elite' consisting of predominantly White male professors, are argued by Gibney to emerge from systemic injustices and biases at every academic stage that perpetuate inequality (Gibney, 2022).

In terms of gender, the figures were higher for women awardees at 30, 32 and 46 per cent respectively; and for people with reported disability the share of awardees had ranged between 1 and 3 per cent for all three roles in this period. Higher award rates were identified among White applicants in all three roles in 2019–20, and individuals who did not report disabilities also tended to have higher award rates. In relation to gender, the picture was more positive. For the first time, in 2019–20, male and female principal investigator applicants enjoyed almost identical award rate (that is, 28 per cent versus 27 per cent, respectively) in the previous six years when applying for principal investigator grants in Biotechnology & Biological Sciences (UKRI, 2021b).

In analysis of monetary value of the awards for UKRI grants in 2019–20, principal investigators and fellows from White ethnicities applied for and received higher award values than minority ethnic applicants. Interestingly, within Biotechnology and Biological Sciences committees, minority ethnic principal investigators had higher award values than their White counterparts in 2019–20. Further analysis may be needed to understand the effect of both disciplines and the nature of funding calls on award values. Still, it is important to emphasise that award rates by value are higher for White principal investigators and White fellow applicants. When sex is taken into account, male principal investigator applicants had a higher award rate by value than female applicants. However, Biotechnology & Biological Sciences committees had the smallest differences in median award values for principal investigators by gender in 2019–20 (UKRI, 2021b). This suggests that there are some positive signs of change for women scientists, and stronger commitment to diversity now underpins UKRI's five-year strategy (UKRI, 2022d).

How research funding bids are assessed and selected has also been put into sharp focus in relation to the representation of minority ethnic panel review members. Success in funding bids—grant capture—can be seen as a seal of excellence, representing a substantial achievement that enhances and secures career positions for those who are awarded such funding. Additionally, participation in peer review processes, funding panels and related activities can provide career advancement because this work comprises citizenship activity seen as a valued and external validation of experience. At the same time, it is crucial that research funding panels are diverse and the processes inclusive for two reasons: first, so that their collective decision making is as free from bias as possible (Ramnani, 2022); and, second, so that more diverse representation at all levels of research is nurtured (Bentley et al, 2017).

Evidence relating to diversity of review panels was provided to the UK House of Commons Science and Technology Select Committee for their Diversity in STEM inquiry. Evidence submitted by Narender Ramnani (2022) included data from 2015–2020 to UKRI's eight research councils. The information returned showed that individuals who disclosed their ethnicity as Asian were underrepresented across all eight research councils, ranging from 3 per cent (Natural Environment) to 7 per cent (Engineering & Physical Sciences). The numbers were even lower for the individuals who disclosed their ethnicity as Black. The highest percentage (2.1) was found in the Economic & Social Research Council, but the percentage ranged between 0.34 per cent and 0.63 per cent for all other research councils in which Black individuals participated. Out of 1,525 committee places in Biotechnology & Biological Sciences, there were no Black committee members over a five-year period (Ramnani, 2022). This historic paucity of diversity on review panels highlights some of the challenges in in efforts to nurture diversity and inclusion in the wider research ecosystem, and it also suggests that the STEM research system needs to work to fulfil its goals of equity in opportunities and outcomes.

A considerable body of knowledge indicates that experiences of STEM vary according to social characteristics. For instance, career experiences have been explored in a study that used survey data of 25,324 STEM professionals in the US across 31 social categories, including gender, ethnicity, sexual identity and disability status (Cech, 2022). The research showed that, compared to other groups, White, able-bodied, heterosexual men experienced systematic advantages that provided them with increased social inclusion, professional respect, opportunities for career enhancement as well as higher salaries. This was in sharp contrast to women identifying as Black, LGBTQ, and disabled who were least likely to experience such privileges, and most likely to experience harassment at work. Differences in social inclusion, harassment and respect could not be clearly explained by variations in family responsibilities, work effort attitudes, job characteristics, background characteristics or human capital, but rather they were due to bias. These results have important implications for equality practitioners: the 'premiums' that White, heterosexual, able-bodied men experience cannot be dismissed as meritocratic rewards for continuous training, work devotion or unique employment circumstances (Cech, 2022).

Implicit bias can affect decision making about shortlisting and funding science: there is no single source of bias in peer review, and potential sources of bias are multiple. Bias can occur on the basis of associations, expectations, cronyism; preference or disfavour for research topics, areas and interests, institutions, individual researchers, gender, ethnicity or professional status; ideas about the constitution of the 'ideal academic' and the 'ideal career'. These biases can affect how panellists 'read, view, hear, question, describe and discuss applications' (Vinkenburg et al, 2022, p. 3; also Fang and Casadevall, 2016). Implications for science and the career success of individual researchers are likely to be numerous and substantial, and implicit bias training as a potential intervention in covered in Chapter 5.

The impact of the COVID-19 pandemic on research careers

The COVID-19 pandemic has adversely affected researchers' long-term career plans. In 2021, the journal *Nature* conducted a Salary and Job Satisfaction Survey[2] with 3,200 self-selected scientists around the world.[3] Findings suggested that, even though most respondents had not been diagnosed with COVID-19, 43 per cent of respondents (equally likely for both men and women) reported that the pandemic had been detrimental to their scientific careers and future prospects, and that the circumstances made it difficult to collect data vital for research and, therefore, careers. Although some participants recognised that they had more chance to focus on writing and conducting literature reviews, most also said that the pandemic meant that they could not discuss ideas with principal investigators, work with internal peers or run their usual laboratory experiments.

Further studies focusing on the impact of the COVID-19 pandemic indicated gender differences. For example, academic mothers felt that their research activities were disrupted during the pandemic, as they had to spend most of their time on teaching and sensed a weakening of their relative competitiveness compared with male and child-free peers (Minello et al, 2021). Another study showed that women's representation as first authors of research articles on COVID-19 was especially low for items published in March and April 2020 (Andersen et al, 2020). These findings align with current thinking that the research productivity of women, especially early-career women, has been more negatively affected by the pandemic than the research productivity of men.

The impact of the COVID-19 pandemic was so great that some researchers felt that their careers had become untenable. The British Neuroscience Association (2020) reported that 25 per cent of respondents were considering leaving the career altogether, and for over 40 per cent of charity-funded early career researchers (ECRs) the main reason given was difficulty in access to funding during the pandemic (Association of

Medical Research Charities, 2020). At the time of writing in 2022, even though some countries had declared 'business as usual' and a return to the pre-pandemic ways of working, EURAXESS UK[4] ran a workshop focusing on the impact of the pandemic on ECRs. This initiative signalled that the long-term impact of the pandemic and associated changes in working practices were still largely unknown. Knowledge about impact is still needed to inform recommendations and mechanisms to support careers in response to limited researcher mobility and the effects of lockdowns, limited collaboration, and network development opportunities (see Chapter 6 for support mechanisms, such as declaration of 'special circumstances').

Summary

- Researchers' abilities to thrive in their careers and feel part of the research ecosystem are crucial for research organisations that rely on excellence for financial stability and reputation.
- Research careers are highly competitive. Current systems and structures tend to provide most benefit to individuals who are well-networked, have access to resources and who 'fit the mould'.
- Diversity, equality and inclusion of researchers from different backgrounds and minority groups is vital to ensure that local and global communities are represented in knowledge-production activities and are served by science to improve health for all.
- Loss of talent from the career 'pipeline' occurs in the context of 'headwinds' and 'tailwinds' that benefit some individuals and create barriers to others. This includes systemic issues of discrimination, bias, stereotyping, availability of and access to resources and networks, among other factors.
- The COVID-19 pandemic has had an adverse and amplifying effect of inequality. These impacts mean that institutions should rethink approaches to how research is conducted, how research careers have been interrupted, and how research and individual research profiles are assessed.

THREE

Why EDI matters
to research organisations

This chapter explores how organisations rationalise and make sense of diversity, specifically the rationales underpinning their actions. For example, EDI is increasingly supported through institutional strategies and visions, in which the diversity of staff populations in academic or research organisations is thought to indicate positive progression. This chapter explores the key reasons for the attention to EDI in organisations, often called the business case and the social justice rationale, which are sometimes thought to work in opposition to one another. Woven through both approaches are glimpses of deeply embedded values within the research system, including the desire to achieve research quality or excellence. Bringing the business and social justice approaches together through a focus on the common good may prove a useful way forward. The chapter concludes with some of the responsibilities that UK research organisations have in relation to EDI.

Why nurturing diversity matters: the case for change

In research contexts, the value and relevance of diversity is increasingly recognised and made visible. This is evidenced by the presence of diversity in the UKRI 2022–27 strategy, in which diversity is one of four principles for change, and in which there is a commitment to 'supporting diversity of ideas, people, activities, skills, institutions and infrastructures' along with 'broadening incentives to avoid homogenisation and promote a diverse portfolio of research and innovation activity in the UK' (UKRI, 2022d, p. 8).

The drive to nurture EDI in organisations, including academic and research institutions—and therefore in careers— can take place for a variety of reasons. Often, the underpinning rationale for attention to EDI in organisations is articulated in terms of either the 'business case' or through appeals to 'social justice'. These two approaches are often seen as opposing one another, and recent scholarship has suggested ways to bring them together or reconcile the approaches.

The business case for diversity and EDI

The original business case for diversity is usually understood as an approach that gained traction in the 1980s, and initially in relation to for-profit contexts. In this approach, workplace or employee diversity was thought to enable organisations to attract and retain talent, better address the needs of diverse customers, and engage in more creative problem-solving (Konrad, 2003). Writing that has supported and bolstered the business case for diversity has included a focus on the relationship between diversity within organisations and their performance and competitiveness (for example, see Cox and Blake, 1991). Fundamentally, business case arguments centre organisational effort to enhance diversity as a 'rational' choice. This late twentieth-century focus on rationality harks back to mid-twentieth-century work, including by economist Gary

Becker, who was an early proponent of the rational view of the firm, in which people's behaviour was thought to be motivated by rational economic choices. When the business case for diversity is reframed for non-profit research contexts (although acknowledging some may be for-profit organisations) diversity enables organisations to attract and sustain a highly talented workforce.

In health and biomedical research, an organisation's purpose, usually laid out in a mission and vision statement, often includes a contribution to the public good through research. Using a business case approach means that work to enhance diversity should help organisations to meet their goals. First, attracting and sustaining a diverse workforce therefore enables organisations to deliver on their mission. As health and biomedical research organisations compete increasingly globally for highly skilled and sometimes scarce talent, broadening the recruitment pool is crucial for sustained success and growth. The presence of a welcoming and diverse workforce helps to support such efforts. Second, if research is to serve diverse populations, the business case suggests that this is best achieved if the workforce is diverse. This means that research teams and research ecosystems are thought to provide the most benefit if the teams and ecosystems are as diverse as the populations that their work serves (or 'representative' of the populations that they seek to serve). However, such approaches would ideally not take advantage of minority status and, as suggested by Sandra Cha and Laura Roberts, an organisation's ability to take advantage of such benefits of diversity is 'ultimately in the hands of its minority employees' (2019, p. 735). Third, and linked to the two previous points, a diverse workforce should be better placed to bring innovative and creative approaches into research. For instance, in research this may include diversity of fields or disciplines as well as diversity of individual characteristics and backgrounds. Innovation, creativity, and productivity in research are sometimes assessed through the lens of research quality, or research excellence, although definitions of research quality are evolving and may vary between fields.

Understanding how organisations change over time is central to understanding how the actions relating to the business case for diversity may become embedded across and between organisations. There are many theories about organisational change. One that focuses on the processes by which organisations start to look and function more like each other was described by Paul DiMaggio and Walter Powell in their article about 'institutional isomorphism'. They describe three mechanisms through which organisations become more similar to one another: 'mimetic' (based on imitation), 'coercive' (based on legislation), and 'normative' (based on shared norms) (DiMaggio and Powell, 1983). These are not necessarily mutually exclusive but provide a useful framework for thinking about how ideas and norms relating to organisational policies and practices may spread. Such institutional change could also be understood through the concept of 'mainstreaming'. In mainstreaming a particular standard is agreed as legitimate in a range of contexts and activities. A standard can be mainstreamed with the support of policies and legislation, for instance as took place in relation to gender across UN initiatives, as described by Emilie Hafner-Burton and Mark Pollack (2002).

Mimetic mechanisms take place when one organisation imitates another and particularly when those within an imitating organisation are uncertain about the potential outcome of their choices (DiMaggio and Powell, 1983). In mimetic processes, people tasked with finding solutions to new challenges (for example, this could include EDI practitioners in academic institutions) search for 'legitimate' and off-the-shelf solutions already practised in reputable and trusted organisations. In the absence of empirical evidence relating to their own organisations, practitioners must hope that if an approach is effective in one setting, then it should work in another. In EDI practice, such diffusion of practice can take place through multiple routes, including the imitation of initiatives taking place in other trusted organisations, through diversity training programmes delivered by consulting firms working between many organisations, or as employees moving between institutions and cross-pollinating practices or by EDI consulting firms. Through 'coercive' mechanisms, organisations become homogenous through compliance with rules and regulations (DiMaggio and Powell, 1983). In the UK, examples include legislative devices, such as the Equality Act 2010 or the Public Sector Equality Duty (briefly described later), which underpin the formulation of human resource (HR) management practices and policies. Lastly, 'normative' mechanisms depend on social duties and obligations, including those that are driven by professional standards (DiMaggio and Powell, 1983). Such homogeneity of practice as a norm means that organisations and institutions find common language and practice through which they build and recognise one another's legitimacy. In the context of EDI more broadly, such commonalities can be fostered by a number of approaches, including Communities of Practice, as described in Chapter 5.

In relation to institutional isomorphism and diversity initiatives, Alain Klarsfeld's work in France provides some clues into how diversity is managed in institutions. This occurs in line with the view that institutions may be motivated to change for reasons unrelated to reaching 'an optimum' in their particular context (Klarsfeld, 2009, p. 323). The author suggests that coercive mechanisms have greater potential to bring about considerable changes in HR management anti-discrimination efforts than voluntary processes. However, coercive, and voluntary practices are intertwined and complement one other.

In higher education (HE) contexts the desire to achieve greater diversity relates to organisational performance and therefore the business case rationale. In particular, internationalisation in HE requires greater diversity in employees and contexts. This is partly driven by, and reflected in, the growth of global university ranking schemes or 'league tables' that amplify difference between organisations and serve to attract national and international talent. International engagement, including that indicated by ratios of international and domestic student numbers, international and domestic staff numbers, and collaborative research partnerships now form part of universities performance indicators (Times Higher Education, 2021). Diversity initiatives can be pursued through internationalisation strategies in academia. A study of 400 European higher education institutions (HEIs) found that embeddedness in a globally competitive academic and research arena meant that internationalisation was seen by universities as a key mechanism for achievement of greater international prestige (Seeber et al, 2016).

Some suggest that the business case for diversity should be applied with caution. This is for several reasons. Diversity-linked benefits to organisational performance and strategic missions will fluctuate depending on temporal, political, geographical, ideological, economic and environmental contexts. As such, there is a risk that diversification of workforces may be an instrumental measure designed to suit certain interests but not fully build long-term vision and planning. Second, 'embracing' diversity for business reasons may benefit some minoritised groups but not all at the same time (Noon, 2007). Examples of such partial attention include the absence of benefit to disabled people (Hewett et al, 2021); the problematic linguistic and action-related homogenisation of Black, Asian, and Minority Ethnic communities under the term BAME that obscures granular difference in equalities (see Chapter 1); and challenges related to the presence and conceptualisations of 'light privilege' within minority ethnic groups, in which a 'light' skin hue may be differentially valued in society (Hargrove, 2019). Third, policies developed on the foundations of business case rationale need to be tailored to contexts. For instance, in their work on ethnic diversity in the UK's National Health Service (NHS), Martin Powell and Nick Johns (2015) conclude that it is 'unwise' to transfer diversity approaches between settings without the development of practices designed for the new context. This is particularly problematic as much work that uses business case models has been carried out in US for-profit contexts. Powell and Johns suggest that for instance these do not immediately translate into the NHS. Equally, it could be suggested that to translate into research and HE, approaches would also need to be tailored to this setting. Fourth, the link between the underpinning rationale for diversity interventions and their outcomes is not straightforward. For instance, an international review of EDI interventions in research and innovation has shown that a reason why interventions failed related to the 'reliance on the business case for diversity: failure to capture the complexity of EDI outcomes in an organisational context' (Moody and Aldercotte, 2019, p. 37).

As such, the focus should move from efforts to increase numerical diversity to efforts that nurture inclusion through relevant policies and practices (Mor Barak et al, 2016). In a report for the WT, Duncan Chambers and colleagues (2017) recommend that this can be achieved with the right implementation of HR policies and practices. These interventions can include strategic recruitment and selection as well as integrated training, development and mentoring packages (for mentoring, see Chapter 5); long-term supported communities and networks (for communities of practice, see Chapter 5); and culture change (for research culture, see Chapter 6).

The caution needed in the application of approaches based on the business case rationale does not undermine the evidence that diversity is associated with novel research, is good for research productivity, and therefore supports research quality. This is demonstrated in work by Yang Yang and colleagues (2022), whose review of 6.6 million medical science research articles across 45 subfields indicated that research articles from teams comprising women and men were more novel and cited more often than articles produced by teams made up of only one gender. Although novelty and citation rates on their own are not necessarily appropriate or responsible indicators of research value or quality, such metrics are relatively easily collected and enable analysis at scale. Understanding why these differences occur remains to be investigated.

The social justice rationale for diversity and EDI

A social justice rationale for diversity foregrounds fairness and thereby the importance of moral dimensions for EDI work. As a widely used concept that has evolved since its origins in nineteenth-century thought, social justice approaches to diversity emphasise the need to address and rectify injustices in wider society. This means that equality is itself a core component of social justice. While the business case for diversity is based on delivery of competitive advantage and performance, the social justice rationale recognises and acts on the basis of a need to address the lived experience of minoritisation and marginalisation (Byrd and Sparkman, 2022).

The benefits of integrating EDI with a social justice case in the research ecosystem include: an overarching effect on researchers and their careers; diversity in research content and analysis; and inclusion in public engagement and knowledge communication. The social justice rationale in health and biomedical research is based on the view that research benefits all people and communities locally, regionally and globally, and that without their engagement, participation and inclusion this goal cannot be accomplished. This approach is highlighted by Amy Bentley and colleagues, who write that attention to diversity within genomic studies may be bolstered when researchers have 'personal connections and interests related to the communities that they aim to serve' (Bentley et al, 2017, p. 258). Diverse researchers can also act as inspiring role models. When children in different parts of the world are asked to imagine a scientist, they tend to evoke laboratories, white lab coats and instruments (Chambers, 1983; Buldu,

2006; Thomson et al, 2019). It is vitally important that future generations of researchers see role models who look and sound like them. This needs, of course, to avoid tokenism and can inspire and signal that everyone belongs in science.

There are myriad ways in which social justice is understood or enacted, which is important because different social contexts and research cultures may have varied approaches. Research culture is discussed in more detail in Chapter 6. For the purposes of this chapter, some consideration of the flexibility and diversity of the social justice area is relevant. Susan Clayton and Susan Opotow (2003) recognise the fluidity and flexibility inherent to social justice, particularly because justice is a product of human action in the context of their societies or settings. This means that each context constructs norms and understandings of social justice and may enact these in certain ways. The authors suggest that individuals and social groups differ in their interpretations of justice as well as in their views about the extent to which justice is more or less important than other propositions, which may include efficiency and effectiveness, pragmatism or wealth creation (Clayton and Opotow, 2003). A country-level example of the differences of this in practice is Michele Moses's (2010) work to compare how France, India, South Africa, the US and Brazil construct rationales for attention to social justice. Moses found that reasons stemming from 'the need to compensate for past discrimination seem to be compelling and salient in India and South Africa, but the longer history in the United States has shown this rationale to be less popularly acceptable and less compelling' (Moses, 2010, p. 218). The consequence is that some groups, individuals and contexts may embrace social justice rationales in different ways, and more readily than others.

Through the lens of the business case, diversification and internationalisation agendas are intertwined and help organisations to meet their purposes in the context of global rankings or league tables. Moves to internationalise universities and research organisations may also relate to ideals of social justice, although these may not always be explicit. For instance, a study of HEIs in Canada, the US and UK discusses internationalisation strategies in relation to students (Buckner et al, 2021). The authors suggest that institutions' use of imagery celebrated diversity, but that race and racism were either 'ignored' or 'externalised' as international. They also found that 'diversity and intercultural awareness' were used to justify student internationalisation.

In another analysis of 32 Canadian HEIs' internationalisation strategies, Elizabeth Buckner and colleagues (2020) found that an appeal and commitment to social justice took place alongside rationales that were focused on business case, particularly relating to research through productivity, reputation and revenue. While the communication of internationalisation in this way may generate support from stakeholders, and while diversity and awareness were given prominence, institutions' strategies contained little in the way of appeals to 'equity, empathy, humility, and civic responsibility' (2020, p. 20).

Social justice arguments for diversity may change as norms in society shift. For example, the tragic events that led to the growth and spread of the Black Lives Matter movement, the disproportionately negative impacts of the COVID-19 pandemic on minoritised groups, the effects of the climate crisis on resource-limited countries, and the warnings about the unravelling of decades of gender equality efforts have ignited renewed debates about nurturing equality, diversity and anti-discrimination in many areas of social policy. It is important to recognise that various social and political global events might affect the pertinence of the social justice rationale and the 'construction' of the urgency to act.

Social justice enables practitioners to approach EDI as the 'right' thing to do to address wide injustices. These interventions can include reverse mentoring programmes to raise awareness of social injustices (for reverse mentoring, see Chapter 5); allyship (see Chapter 5); and targeted and positive action interventions to level up the playing field (see Chapter 6).

It seems reasonable to suggest that individuals and groups engaged in EDI action who are thinking about their work through social justice may be particularly and appropriately attached to and motivated by their work. It may be important to be prepared for the fluctuating peaks and troughs of interest, the openness to change, resistance, and the availability of resources. Finally, it can be reasonably argued that all the attention to diversity and EDI can be conceptualised through a social justice lens. Understanding how organisations may or may not appeal to social justice rationale is, in itself, interesting and provides an open window into how organisations work.

Bringing together the business case and social justice rationale

The business case and social justice rationale can seem opposed and contradictory, with one approach emphasising productivity and the other focused on fairness. Helpfully, several scholars have considered how to reconcile the two, including those working from organisational ethics and human relations perspectives. This can form the basis for a view of EDI in health and biomedical research that can reasonably refer to both the business case and the social justice rationale.

Hans van Dijk and colleagues (2012) suggest that the 'stalemate' between the business case and the social justice rationale (the latter described as the 'equality' perspective) can be resolved through the application of virtue ethics and values in HR work. Attending to virtues, for instance in recruitment, focuses attention on the qualities that enables people to do a particular role and helps to reduce prejudice. While work with virtues focuses on individuals, including relationships between them, a focus on values attends to the guiding principles of organisations, which help to define the relative importance of virtues. The authors suggest that values are usually core to organisational strategy. For clarity, organisational strategies often include a set of values that encapsulate ways of working. The values proposed by van Dijk et al are instead the heart of organisational strategy and these are usually reflected in vision and mission statements. Similarly, from an organisational management perspective, George Gotsis and Zoe Kortisi (2013) draw on philosophy, particularly ethics, to highlight the potential of an 'ethic of care' approach to diversity management in organisations. In this, people and their dignity are the foundation of organisational diversity management. Finally, Sandrine Frémeaux builds on the work of scholars focused on ethics. Through this, Frémeaux identifies the 'common good' as a way of conceptualising diversity through its value and benefit to individuals and the 'flourishing of individuals is based on the flourishing of the community' (Frémeaux, 2020, p. 200). This conceptualisation of diversity management in organisations provides an alternative to approaches that might make diversity endeavours transactional or focused on objectives other than humanity's wellbeing. In other words, a good context enables people to thrive.

Taking a human relations approach, Marilyn Byrd and Torrence Sparkman also emphasise the centrality of people and propose that action-oriented work can reconcile differences between the approaches to diversity (Byrd and Sparkman, 2022). They suggest that the business case for diversity has now incorporated further areas that organisations need to consider to perform well. These include the diversity of ideas, cognitive skills, creativity and talent, abilities, and perspectives and they suggest that improving and building relationships among people enables challenge to systems through micro-level interventions.

Drawing on the collective experience of the pandemic and its impact on the research ecosystem, Gemma Derrick (2020) reflects on how the crisis appeared to trigger a 'kinder' research culture at the time, as researchers, academics and publishers chose to communicate and collaborate with each other with empathy and respect. For many working in health and biomedicine, the COVID-19 pandemic was a time of intense research activity, as groups and individuals refocused onto the disease, including prevention detection, treatment and understanding numerous impacts. The collaborative nature of research was more evident than ever, alongside intense and rapid design of studies and the growth in a culture of pre-print publication. While in Chapter 2 the detrimental impact of the pandemic is discussed, it could also be argued that a 'kinder' approach that took note of circumstances was vital to the research ecosystem's ability to deliver research at pace and of value.

Insights from organisational ethics and human relations provide ways of thinking about EDI that acknowledge the relevance of the business and social justice rationales while also offering overarching perspectives and principles. In health and biomedical research organisations, the business case and social justice rationale may function alongside one another to offer a legitimate and compelling basis for EDI work. To bridge the gap between these approaches, an emphasis on the virtues, value, common good and dignified, respectful ways of working may already align with the purpose of research and therefore research organisations.

At root, research organisations often frame their purposes in relation to the provision of benefit to wider society, for instance through new understanding of, prevention measures, or treatment for health and ill health. While the business case's guiding principle is that diversity enhances productivity and the social justice rationale suggests that EDI is needed for principles of fairness, together these enable organisations to meet their purposes in relation to wider society. At least, this is the case if productivity in health and biomedical research is understood in terms of benefit to health and wellbeing, with any profit or surplus—where relevant—seen as a side-effect rather than core aim of the work. The achievement of benefit for and in society is the common good, which is usually reflected in the articulation of the mission and vision of research organisations (Figure 3.1).

Figure 3.1: Complementary rationales for diversity in health and biomedical research

Organisations may do well to use approaches and methods that enable the business case and social justice rationale to be both foundational and visible. These do not need to be seen as mutually exclusive, opposed or in tension, but they can instead be viewed through a single frame of reference—the common good—that is already so often articulated in institutional and organisational strategies. This can underpin comprehensive packages of interventions to support careers to thrive in fair and inclusive environments that also achieve organisations' purposes.

What are the responsibilities of public bodies and universities?

In the UK, legislation that covers employment, education and equality encompasses the research ecosystem. In addition, broader legislation influences the measures and provisions that take place in all parts of society. In Chapter 1, the requirements of the Equality Act 2010 relating to 'protected characteristics' were described. These characteristics are often named and therefore visible in individual organisations' policies and practice. Similarly, in relation to gender, gender pay gap regulations apply, which specify that women and men must receive equal pay for equal work. In addition, gender pay gap reporting regulations started in 2017, which requires that certain organisations must report gender pay gap information on an annual basis. This reporting requirement applies to most research organisations such as research-performing HE contexts. Also, as HEIs are public bodies in England, Wales and Scotland, the Public Sector Equality Duty (PSED) applies. The PSED requires public bodies to address and remove discrimination and advance equality of opportunity: this includes in strategic as well as operational decisions.

All legislation has the effect of bringing about change, which is a 'coercive' mechanism for institutional isomorphism in DiMaggio and Powell's theory presented earlier (1983). Although probably less well known than equality and gender pay legislation, the PSED is particularly focused on generating change. The PSED contains three specific aims: the first focuses on the elimination of discrimination, harassment, victimisation and any other prohibited conduct. The second and third focus on advancement of equality of opportunity and fostering good relations between people who share a relevant protected characteristic and people who do not share it. Research organisations often conduct considerable visible work to advance opportunity and foster good relations. Work to eliminate discrimination and other prohibited conduct may be considerable but less visible if such work takes place after an occurrence of misconduct rather than in efforts to prevent it.

It is suggested that, in practice, the three aims of PSED could be achieved through:

removing or minimising disadvantages suffered by people due to their protected characteristics; taking steps to meet the needs of people from protected groups where these are different from the needs of other people; and encouraging people from protected groups to participate in public life or in other activities where their participation is disproportionately low. (EHRC, 2022, np)

It is worth noting that the PSED does not oblige public bodies to achieve specific targets or changes. However, they must consider how their ways of working might impact those with protected characteristics, and 'due regard' must be fulfilled consciously at every stage of policy-making (EHRC, 2022). The PSED works to 'stimulate self-regulation at an organisational level through prompting local actors to deliberate and use their know-how to "translate" the requirements of the PSED according to their own organisational requirements' (Manfredi et al, 2019, p. 4).

Although the PSED approach has been criticised for being 'soft touch' (Fredman, 2014), Simonetta Manfredi et al's (2018) legal scholarship review suggests that the PSED does nudge organisations to embed equality standards and interventions. This finding also aligns with the previously mentioned work on the effectiveness of coercive approaches by Klarsfeld (2009). Furthermore, the participatory elements in the processes of self-regulation that are underpinned by PSED can themselves foster equality outcomes when such participation includes members of vulnerable groups (Manfredi et al, 2018).

However, it is argued that even if organisations succeed in ensuring equality in policy mechanisms, the lived experience of equality, equity and inclusion is not complete (Özbilgin, 2009). For instance, Kalwant Bhopal (2020) acknowledges some progress has been made in the supporting of careers of Black and minority ethnic women, but most of the support tends to benefit White women. Moreover, there is the risk that equality efforts become a tick-box exercise of 'doing the document rather than doing the doing' (Ahmed, 2007, p. 599). Therefore, any practices and policies that aim to address inequalities must be included within the 'cultural organisation of institutions' and be underpinned by key objectives that require accountability (Bhopal, 2020, p. 707). As argued by Colin Scott (2020), a commitment to EDI should mean that academic institutions change how they operate at every level. This transformation should apply not only to the subjects of academic institutions, such as students, researchers, academics, professional staff and leaders. Scott's work suggests that this applies to transversal activities, processes and practices within academic institutions. These activities relate to how learning takes place; who the learners are; who does research, whom and what research focuses on, how research is done, who benefits from research; how relationships are organised and how enterprise is leveraged; and how partnership and public engagement is delivered (Scott, 2020).

Summary

- Activities to nurture EDI may be underpinned by either or both the business case and social justice rationale.
- The business case and social justice rationale do not have to be seen as mutually exclusive.
- Approaching EDI in health and biomedical research through a common good perspective can be a good match to organisational strategy and purpose.
- Approaches such as legislation and imposed duties can influence the motivation of institutions to nurture EDI and should be combined with voluntary approaches.
- Nurturing diversity requires work to take stock and transform the wider cultural context. This helps all individuals feel included and thrive.
- EDI needs to be embedded in strategy and activities, processes and practices that flow from strategic goals.

FOUR

Why EDI matters to individual researchers and researcher communities

In Chapter 3, we discussed how organisations rationalise and make sense of diversity and suggested how different rationales can be combined to create a compelling and sustainable case for diversity. This chapter outlines selected literature relating to the lived experiences of minoritised and disadvantaged groups of researchers and academics. In so doing, the chapter seeks to amplify those experiences that are often shared with such emotion and are key to understanding why EDI action is so vital and matters to individuals.

Lived experience as a legitimate source of knowledge

The capacity to thrive professionally in an autonomous and creative way is underpinned by an individual's agency: the ability and power to act and to be the architect of one's own life. However, whether individuals can meet their potential and have agency is always shaped by circumstances and context. Individuals experience the invisible but present structures of social inequality, racism, power and order hierarchies (vertical and horizontal segregation), patriarchy, heteronormativity (norms related to sexual orientation), cisnormativity (norms related to gender identity matching one's sex at birth), chrononormativity (norms related to age), ableism (norms related to being able-bodied), and dominant political ideologies, and the discrimination and disadvantage through limited opportunities that these structures create.

We acknowledge and accept that we are unable to include everyone's voices in this chapter, and that matters relating to race, ethnicity, gender and disability are the main threads. The focus shifts from less personal and often anonymous evidence to narratives about daily challenges and experiences of marginalisation, inequality, loneliness and exclusion. Surfacing lived experiences recognises discrimination and its impact (Tate, 2017) and acknowledges that such experiences are legitimate sources of knowledge (Collins, 1990; Hartsock, 1998; Parker and Lynn, 2002). The empirical evidence that follows focuses on STEM (science, technology, engineering, mathematics), health, and biomedical academic and research stakeholders where this is available but draws from other discipline contexts where applicable.

Inhabiting 'White' spaces

UK higher education (HE) spaces are disproportionately White. Recent research and writing by Black academics and colleagues shine a light on the challenges that this presents.

Individual pathways into research careers usually, although not always, start with university education. Representation of Black, Asian and minority ethnic groups among first degree undergraduates and taught postgraduates is slightly lower than for White populations (27 per cent and 24 per cent, respectively, Advance HE, 2021b). However, these figures mask the 'inequitable experiences of those from marginalised groups of all ages who do "get in", as they have to formulate adaptive tactics to "get by" in the institution before they can attend to "getting on"' (Oman et al, 2015, np). For example, White students are much more likely to be awarded a first or upper-second class degree than students from Black, Asian, and mixed ethnicity backgrounds. This award gap between student groups was 13 per cent in 2017–18 (UUK and NUS, 2019). As Shirley Anne Tate and Paul Bagguley (2017, p. 292) explain: '[s]uch an attainment gap should make universities ponder what it is about, what happens within walls, classrooms, and curricula that supresses the emergence of BME student excellence'. Tate also argues that some experiences are 'unvoicable' within institutions, and specifically, '[q]uestions about and answers that voice the daily racism and racist micro-aggressions' are particularly silenced (Tate, 2017, p. 54). Both authors draw attention to the accounts of stakeholders who report the lived experiences of 'continuing institutional racism, curricula which continue to be Euro-centric and faculty which do not reflect the UK's demographic diversity' (Tate and Bagguley, 2017, p. 292). Kalwant Bhopal (2014) suggests that, despite progress in race equality in the UK HE in recent years, students identifying as Black Caribbean particularly— and generally students from minority ethnic backgrounds— continue to experience disparities including in subsequent career progression. Reasons for such challenges to progression are complex and include the likelihood that Black students are less likely to have attended the prestigious, pre-1992, Russell Group universities (Bhopal, 2014). This entrenches divisions based on social class, social mobility and access to powerful and influential networks (Bhopal, 2014).

Racially-motivated discrimination and prejudice have profound negative impacts, and the regularity and constancy of such experiences compounds this (Wallace et al, 2016). Experiences involving racism cause harm to those who experience racism directly, and also to those who witness these occurrences. For instance, Kimberly Truong and colleagues (2016) highlight the ways in which race and racism can shape the experiences of minority ethnic doctoral students studying a range of disciplines including biomedical sciences, biostatistics and public health in the US. In their qualitative study with doctoral students, they examine students' experiences of the 'vicarious racism'. Vicarious racism takes place when racism is ostensibly targeted at another person or group, but is witnessed by an individual and therefore impacts on that individual too. These experiences had negative emotional and psychological effects, including frustration, anger and sadness (Truong et al, 2016).

People who live with intersecting identities relating to their gender, ethnicity and class can experience additional challenges, isolation and stigma. A striking demonstration of these experiences is provided in work by Deidre Bowen (2012), who conducted a study that included over 330 individuals attending a biomedical research conference for minority ethnic students in the US. The students were all in receipt of mentoring programmes that aimed to nurture their careers. The findings indicated that, while ethnicity was a salient aspect of experiences in HE, gendered differences were apparent. For instance, while Hispanic/Latinx men and women reported receiving similar levels of encouragement from faculty members, African American women reported the lowest rate of faculty encouragement to speak about their career aspirations compared with their African American male and Hispanic/Latinx female and male counterparts. When students were asked whether they thought that 'minority students could only get accepted to college with the help of positive action', over 35 per cent of women, compared with 23 per cent of men, responded that both faculty (staff) and students felt that way. This suggest that individuals with intersecting identities of being a woman and African American at the same time were more likely to identify barriers to access to HE.

The context in which students learn is vital to action that seeks to equalise research career outcomes. Equally, the content of the curricula is important, with moves to create more inclusive learning experiences gaining pace in HE. For instance, Sarah Wong and colleagues describe endeavours to decolonise learning materials at UCL Medical School in London, UK and highlight earlier progress in decolonisation within arts, humanities and social sciences subjects (Wong et al, 2021). This work resonates with wider and growing concern that there is a need to address structural racism in medical practice (Crear-Perry et al, 2020). Although deep exploration of curricula is beyond the scope of this book, it is important to be mindful that the learning experiences that future researchers have can pave the way for career diversity. In turn, an inclusive learning environment lays the groundwork for the content of the research and the idea that diversity in research settings is normal rather than exceptional.

As described by Jason Arday, a visible lack of academics from minority ethnic backgrounds reminds research students considering their next career steps that entry into academia and its opportunities remain challenging and a lonely endeavour (2021). Underrepresentation of minority ethnic communities continues to characterise academia. In the UK in 2020/21, only 155 out of 22,795 (0.67 per cent) professors were Black (defined as Black/ African/Caribbean/Black British), and there were only 40 (0.17 per cent) Black women professors (Advance HE, 2021a). In the US, numbers are not much better: in 2020, full-time faculty in degree-granting post-secondary institutions comprised 188,692 tenured associate and full professors. Of those, 3,186 (1.7 per cent) were Black women, and 2,982 (1.6 per cent) Hispanic women (National Center for Education Statistics, 2021).

Narratives about experiences of minoritisation demonstrate the realities of not being accepted, not being included, disconnection and invisibility. However, as Arday (2022, p. 82) describes, there is still a persistent 'dearth of research on the lived racialised experiences of Black people particularly within the Academy'. Evidence that does exist includes narratives from Black female academics who share their career stories. As highlighted by Marcia Wilson, these provide a means to interrogate and recognise 'racial storms in the academy'

(Wilson, 2017, p. 119). Reflective accounts are provided by Victoria Showunmi and Uvanney Maylor (2013), and by Deborah Gabriel and Shirley Anne Tate (2017), which describe experiences of indignity, separation and exclusion in accounts about experiences of Black female professors (Jackson, 2017). In such accounts Black women recount how their credibility is questioned, how they are framed as not having the right 'potential', how they are described as not yet 'ready' for progression, and myriad other subtle and overt experiences that serve to undermine and inhibit careers.

In representations of experiences of minoritised academics, some experiences are framed as 'microaggressions'. This term is not meant to imply that some actions are not significant, more that aggressions take place frequently in such a manner that they are ever present, build, and cause considerable harm over a period of time. Work by Jiona Lewis and colleagues indicates that experiencing microaggressions affects students' sense of belonging, and that microaggressions are 'rooted in structural and systemic racism experienced by people of color' (Lewis et al, 2021, p. 1053). Derald Wing Sue and colleagues (2007) identify three forms of such indignities: microassaults, microinsults and microinvalidations. All of these may be enacted by people who might not even be aware of the impact of their actions, or enacted with clear intentions. These experiences of indignity and feelings of separation and exclusion are echoed in accounts of Black female professors (Jackson, 2017; Wilson, 2017). At the same time, we acknowledge Emma Dabiri's contribution that an excessive focus on 'microaggressions' can 'obscure, distract and divert attention away from policies, legislation and institutions' (2021, p. 87) that (re)produce inequality and disadvantage, which can only be fully addressed with a systemic change.

A recent inquiry into racial harassment in UK universities by the UK's Equality and Human Rights Commission (EHRC) (2019) indicated that racial harassment is a considerable issue and that most people affected did not report the harassment. When complaints were made, many were not upheld. Fewer than four in ten student complaints, and around one in six staff complaints were upheld and offered some redress. According to the report, around one in three (38 per cent) most recent university staff complaints concerning racial harassment had been investigated and not upheld; 5 per cent were unresolved due to a lack of evidence, and 8 per cent were withdrawn (EHRC, 2019). Importantly, evidence collected in the inquiry indicated that universities often did not feel able to tell the complainant when sanctions had taken place. This was because universities had concerns about adherence to data protection regulations. Individuals felt that support was not in place and, importantly, the lack of wider visibility about the outcomes of complaints about racial harassment is a 'missed opportunity' to make the consequences of harassment visible to the wider community. Such visibility could have positive benefit and reduce incidence, but also encourage colleagues to report racial harassment (EHRC, 2019).

Writings about race and ethnicity tend to focus on academia widely and across disciplines and include exploration of issues in academic career, progression and belonging. It is reasonable to infer that the experiences are not unique to particular fields and, as such, individuals working in health and biomedicine are likely to have similar experiences to those in other areas. How can EDI be nurtured in settings where colleagues have such experiences? The first step may be to build awareness of the scale, depth and detail of exclusion and discrimination. Reading original versions of the accounts of minoritised people are good starting points, as summarising them here cannot adequately convey their content. Chapters 5 and 6 provide strategies that might help to begin to address some of the challenges discussed here through individual, community and system-based interventions.

Inhabiting gendered spaces

Among the intake of health and biomedical undergraduate students, gender equality has largely been achieved and, in some areas, there are proportionately more female than male students. However, considerable evidence points to disparities as careers progress and the description of the 'leaky pipeline' (Chapter 2) pointed towards some of the issues relating to retention of skilled individuals. Of these, considerations relating to gender have been described in particular detail, with research often pointing towards deeply entrenched norms and expectations of gendered roles.

Work to understand women's experiences in research and academic roles include a focus on under-representation at senior levels spanning many fields and disciplines (see, for instance, Ashencaen Crabtree et al, 2017). Understanding women's access to resources for research is important: for instance, a UK survey of 365 individuals—of whom 83 per cent worked in life sciences—indicated that women principal investigators received lower salaries and had access to less space and fewer staff members to support their research than their male peers (Acton et al, 2019). During the COVID-19 pandemic, the reasons for some disparities came into sharp relief, with early commentaries indicating that women's publication rates fell behind those of men, as women took on additional childcare work (Viglione, 2020). Despite a plethora of initiatives designed to enable flexible work arrangements and a work–life balance, many women say that a culture and expected scale of research productivity that relies on long hours is simply not compatible with responsibilities outside work.

An interview study participant reflected on how he found that women were often not able to accept invitations to speak at his institution because of caring responsibilities. In response, he was putting in place a budget to help, in the knowledge that when women could not travel to give talks this impacted on their careers:

'Especially during the Covid pandemic or even after a little bit when you ask a highly achieving research woman to come [to visit], you pay [for] everything and she's completely happy to come, but 90% of the times 'I can't go because I have some parents to take care of, I have a child'—and always a woman. I've never found a man telling me 'I'm sorry, I cannot travel because I have a child'. And this is something that I'm trying to help the most. And I'm now trying to allocate some budget for travelling with kids, for speakers [...] this is something that institution is supporting me [with].' (Participant 1, man, researcher, EDI practitioner)

The ideal of researcher productivity based on long working hours also impacts on men. The same participant shared his story, and described a desire to live a full life alongside his research career. He emphasised his growing awareness that this meant that he was unable to 'compete' with colleagues:

'Well, I'm a father of three kids [...] with my first daughter I started to realise that I could not compete with my colleagues, even as a male researcher [...]. There's no timetable, you don't finish [...] I wanted to be father, I wanted to be dad, and I wanted to get home to my kids, and wanted to be home with my wife.' (Participant 1, man, researcher, EDI practitioner)

Intergenerational differences in how women experience research careers and institutional cultures also play an important role. An EDI practitioner with a background in science reflected on resistance affecting young female researchers:

'[A]t a personal level it's very rewarding to see that we are making progress and that we are able to even if it's really small, make a difference. But funnily enough, there's still a lot of lack of awareness, denial, and resistances and even in older women, full professors who maybe were like the only one in their class back in the day and they somehow don't like that young women scientists have it

'easier' than they did, so […] for me it was so difficult.'
(Participant 4, woman, EDI practitioner)

Clinical academic research is one area in which studies have tried to understand ongoing gender disparities. The evidence that expectations and workplace cultures is gendered has a global relevance. For example, work by Anna Heffron and colleagues with postgraduate students in the US indicates that women students were impacted by expectations, with one male participant saying that they felt that 'women are not given the credit and clout they deserve, both in science and medicine' and another saying 'I frequently observe female residents and students being called nurses' (Heffron et al, 2021, p. 95). In UK-based work, it has been shown that women still leave their careers in academic medicine because of gender-based considerations. For instance, Victoria Salem and colleagues (2022) conducted a focus group with clinically qualified women to find out why they had decided not to pursue academic (and therefore research) careers after completion of their research degrees. As recently as 2018, women were saying that influences included work–life balance, pregnancy and maternity as reasons for them not to take the next step into clinical academia. In the large London medical school from which the women had graduated, the spaces themselves were gendered, as a participant recalled: 'one changing room has a male sign on it and says doctors. The female changing rooms says nurses' (Salem et al, 2022, p. 6). Focus group participants also indicated that the workplace culture was built on full-time uninterrupted availability. This was thought to be based on norms that men were more likely to fit into, while women thought that they were less likely to be available in this way.

Institutional ableism

When organisations and institutions deliver their work based on ableist assumptions, this means that they do not consider the varying extent of ability to access and use of resources and systems. This approach disregards individuals, creates exclusion and places barriers in the way of research careers. Even in contexts where awareness raising and education about ableism takes place, it is crucial to amplify and listen to people who live with the experience of disability or neurodiversity, and further bridge this experience with colleagues who are neurotypical or not disabled, for mutual understanding. This is emphasised by an autistic participant interviewed for the study:

'But I go back to my point of lived experience because actually that is vital when it comes to, certainly EDI, in my humble opinion. Yes, we could have all the facts going and that's brilliant but actually, what is it like for you? I also turn it around and go, actually we don't have the questioning from the opposite side. Actually, what is it like for you as a researcher, or you as my colleague to work with me? How do you find it? How do I come across to you? How can I improve or try to or change my philosophy to that as best as I can, knowing full well that change can be problematic.

I've got to really, really concentrate and it's very draining [...] if you shout across the office and I go 'oi' really, really quickly, you know, if you shout really loudly, it's distractive. Yes. Now imagine that every five minutes or every ten minutes [...] your anxiety levels probably shoot through the roof. Constantly on edge, constantly going 'what the hell's going on?' [...] sensory sensitivity is all these different things coming to play.' (Participant 6, man, independent researcher, EDI advocate and practitioner)

Individuals with disabilities can be highly visible or may also be completely invisible if their disability is not physically apparent, if they are neurodivergent, or their condition is undisclosed to a broader team. This can be problematic in relation to the unintended consequences of colleagues' behaviour and activity on the individual's wellbeing. Invisible conditions may also fuel perceptions that one's disability is not properly validated and that minority status is unacknowledged (Bannerman et al, 2020).

The number of students declaring disability has recently increased. Around 15 per cent of students studying on research and taught postgraduate programmes declare disabilities (Advance HE, 2021b), which compares with 21 per cent of the general UK population declaring disability (Office for National Statistics, 2022). The proportion of students declaring disability is relatively high compared with 4.5 per cent research-only staff and 5.2 per cent of research and teaching staff in HE declaring disability (HESA, 2022d). However, the percentages of students with a declared disability in subjects such as Medicine and Dentistry at a research postgraduate level stand at 4.7 per cent, which represents a considerable underrepresentation compared with subjects such as Physical Sciences (9.8 per cent), Social Studies (9.7 per cent) and Biological Sciences (9.1 per cent) (Advance HE, 2021b). Rigorous empirical studies that attend to experiences of disabled PhD students and research staff are needed to underpin appropriate mechanisms that can nurture research careers on staff members' own terms.

Writing from their lived experience, a neuroscientist with motor neurone disease, Justin Yerbury, and a psychologist, Rachel Yerbury (2021), reflect on the Australian context, and reasons for the disability gap between undergraduate students and staff populations. They identify several factors: institutional ableism, expectations of high levels of output to secure grants, elitism and exclusion prevalent in science, a lack of infrastructural and policy support, and physical, psychological and emotional tolls on individual disabled researchers. Institutional or team-level climate and culture may prevent disabled researchers from feeling confident and comfortable to disclose their condition for fear of being misunderstood, undervalued, discriminated or stigmatised. Yerbury and

Yerbury argue that the medical model of disability (as briefly presented in Chapter 1) is partially responsible for the view of disability as 'a liability, risk, and disadvantage, rather than seeing the diversity and lived experience it brings as an asset' (2021, p. 508). In the UK, Stephanie Hannam-Swain (2017) provides a narrative account of the challenges she encountered as a disabled PhD student framed as the 'additional labour' impacting on her studies. In Australia, an interview-based study with six PhD candidates in social and health sciences by Joshua Spier and Kristin Natalier (2021) highlights how the failure to provide reasonable adjustments was distressing and the cost to them as PhD candidates was sizeable. All these accounts indicate the need to seek and acknowledge experiences to enable institutions to understand current gaps in their provision for disabled students and colleagues, including people who are not yet adequately represented in the workforce.

Implications for mental health and wellbeing

Experiences of exclusion and minoritisation can have adverse consequences for individuals. Some of these are described above, and this section brings some additional impacts into view. Efforts to fit in and the toll these experiences take has been explored by Laura Hull and colleagues (2017) in their study with individuals with autism spectrum conditions. Participants used masking or camouflaging strategies in social situations, which took considerable energy to sustain. Such attempts at fitting in could lead to anxiety, exhaustion, and questioning self-identity. In the study interviews, participant 6 described similar experience and the 'upset' this caused him:

'It's that understanding that if I'm completely off my baseline, then I'm not going to be good, I'm not going to be brilliant [...] So I had a slight incident where everybody thought I was relatively okay. I wasn't. I masked completely and utterly [...] I was upset for

a good four days because I'm trying to work it out.'
(Participant 6, man, independent researcher, EDI
advocate and practitioner)

Experiences of racial discrimination also have an impact
on the mental health of people from minority ethnic
backgrounds (Ashe et al, 2019). Critical events, such as social
unrest in response to the George Floyd case coinciding with
the COVID-19 pandemic and its disproportionate impact
on communities of colour, have amplified 'racial trauma'
(White et al, 2021). For students, identifying as a member of
a sexual minority group or being female within a university
setting is also associated with mental health outcomes (Balloo
et al, 2022). One of the barriers for seeking help for mental
health support is cultural differentiation in how people make
sense of mental health. Research has shown that there exists
stigmatisation of mental ill health in families of 'African,
Caribbean and similar ethnicity' backgrounds and the ensuing
'silencing' of the issue (Dare et al, 2022). This may be one of
the reasons why mental health difficulties are less likely to be
declared by Asian and Black individuals (Balloo et al, 2022).
This finding echoes a study by Myles-Jay Linton and colleagues
(2022) about whether and why students decide to permit their
university to notify their emergency contacts if they develop
serious mental health concerns. The authors found that Black,
Asian and minority ethnic students were less likely to permit
universities to do so than White students; and students with
existing mental health issues were also less likely to opt into
this arrangement. The two most frequent reasons given by
students for not opting in were concerns about not wishing
to 'worry' emergency contacts and preferring to deal with the
situation on their own (Linton et al, 2022).

However, even if the barrier to help-seeking is overcome, the
situation for minority ethnic students and staff is exacerbated by
the limited access to culturally appropriate mental health services.
Arday (2018) examined the negative impacts of inequality and

discrimination among minority ethnic university students; these included the effects on mental health and highlighted the paucity of 'access to culturally appropriate services that are cognizant of the racialized plights' (Arday, 2018, p. 1). Arday's work shows that university as a predominantly 'white' space triggers anxiety and stress in minority ethnic students; and how negative stereotypes mispresented minority ethnic communities as anti-social, unemployed and unemployable, and reliant on state welfare (Arday, 2018). Studies with minority ethnic academic staff show similar contours in mental health experiences, highlighting the presence of stigma, isolation, exclusion and victimisation. These occur in the context of the absence of awareness by mental health support staff about the scale and scope of racism experienced by their academic colleagues (Arday, 2022). For instance, Arday's work highlights that self-organised 'safe spaces' to share mental health difficulties run for and by minority ethnic staff sometimes amplified and worsened negative experiences by creating 'toxic echo chambers' that caused people to relive their trauma. Such studies and the compelling individual accounts emphasise an urgent need to build adequate response mechanisms. For example, recommendations include the provision of access to practitioners who are adequately trained or bring lived experiences as minority ethnic background themselves, and a range of improvements that should nurture 'culturally sensitive interactions and services' (Arday, 2018, p. 21).

International students, or those who work away from their home countries, may encounter unique challenges in relation to dimensions of diversity and any need for support for mental health. Kaite Koo and Gudrun Nyunt (2020) have developed a framework for culturally sensitive mental health assessment. Developed for international students, the work has broader applicability as it is designed for academic and support staff as well as healthcare professionals. The framework incorporates a deep appreciation of specific race, ethnic and cultural differences; attitudes, beliefs, knowledge and skills to deliver effective mental health services; and the location in which these layers of competence are needed: both within practitioner and institution. The authors argue that such synergy between micro- and meso-level interventions is needed for interventions and services to be culturally sensitive and effective (Koo and Nyunt, 2020).

Although this section has focused on the mental health impacts of minoritisation, research cultures can seek to include and foster wellbeing in the context of careers. In Chapter 6, the role of group approaches and the importance of communities is described.

Summary

- This chapter has explored experiences in relation to some key dimensions of diversity. Framing the chapter in this way has intended to highlight a selection of current challenges, although such challenges are invariably experienced in an intersectional way.
- Lived experience is a legitimate source of knowledge that complements numerical data. It is crucial to listen to the marginalised and minoritised voices to enable a reflexive and thorough appraisal of EDI in academic and research organisations. This will in turn enable targeted and effective interventions.
- Evidence suggests that some individuals perceive and experience research and academic settings as unwelcoming, biased and incompatible with their needs. Much work is required to attend to these issues through carefully designed interventions (see Chapters 5 and 6).
- Research and academic environments that do not adequately nurture diversity and equality pose risks to minority and underrepresented groups, as well as the broader community.
- Impacts can include experiences of isolation, discrimination, microaggression and marginalisation and mental health difficulties, which in turn impact on the individuals' prospects and their ability to thrive in research career.

FIVE

How can EDI be nurtured through communities and individuals?

This chapter describes interventions to support individuals and research communities in their pursuit of careers and professional development as researchers. Avoiding the deficit model, Chapter 5 focuses on how change and equality agents within organisations can put measures in place that make a difference to individuals constrained by the legacy of wider social forces that (re)produce inequalities. In Chapter 4, evidence about the lived experience of inequality and marginalisation was included to sensitise readers to potential challenges and barriers that individuals may face. This lays the foundations for interventions presented here and in Chapter 6, in which quotations from interview participants give depth and voice.

Supporting communities and individuals

To nurture EDI, research organisations can deliver interventions at various levels. At a collective level, organisations can offer resources for mechanisms that support and enable community building, networks, groups, peer mentoring and other grassroots initiatives. Organisations can also provide direct help to individuals, for instance through tailored interventions that nurture equity such as positive action schemes (see Chapter 6).

Interventions that support individuals and communities may help to develop awareness of EDI and facilitate systemic changes. However, these approaches should be understood as only one type of tool amid many others that may help to bring about wider change. Approaches that support researchers by asking them to change may reinforce inequalities through the focus on 'fixing' individuals rather than repairing the broader system. When an intervention seeks to change an individual's beliefs or behaviours, the implication is that there is a deficit in the individual that can be addressed as part of a professional self-improvement journey. It does not account for structures and circumstances that may prevent individuals from having equal access to myriad opportunities including education and societal expectations. Approaches that are only based on self-improvement do not account for challenges faced by minoritised and underrepresented groups experiencing systemic barriers as well as racism, bias, microaggressions, discrimination and inaction. Such barriers prevent people from reaching their career and professional goals, as well as impact on beliefs about whether certain goals are within their reach. As such, examples of interventions presented in this chapter need to form part of a wider EDI programme that combines community and individual approaches with wider system changes presented in Chapter 6.

Those who deliver EDI work within organisations are change agents, who may actively embrace and talk about their roles or who may simply embody EDI practice in their everyday work. Regardless of approach, change agents may reflect on their practice, and consider with reflexivity how their own position affects the work that they do. Reflection enables those involved to consider how to move forward and put changes into policy or practice. Reflexivity enables a person to stand back from their usual way of thinking and to critique how that policy or practice may affect others, why it is or is not chosen, or prioritised, and whether it brings about the desired change. Ways to achieve this might include formal equality impact assessment, ongoing discussions with members of research communities, or reflective questioning of their own practice.

There are many ways of fostering change through support for communities and individuals. In health and biomedical research contexts, key approaches have been mentoring, communities of practice, allyship and implicit bias training. Others exist, but these are the most common and widely known about. Their history and evidence about their value are discussed, and reflection is offered.

Mentoring and mentorship

Mentoring and mentorship usually, although not always, involve formal or informal pairing between an individual and a more experienced colleague. There are other approaches, discussed later, that include reverse mentoring and group-based methods. Mentoring provides access to support and experience in relation to work, career development and related matters. Although people in workplaces have always sought to learn from one another, mentoring with particular emphasis on careers and equity gained traction in the 1990s and is now a well-established part of many research environments—so much so that staff may expect to receive mentorship or to be asked to provide it. Formalisation of mentorship and its inclusion in career development plans signal that organisations have realised this intervention can offer positive outcomes for learning, advancement and productivity.

Mentoring, mentoring circles, e-mentoring and one-to-one support programmes are recommended as approaches that can boost the prospects of underrepresented groups in academia generally (Mahayosnand et al, 2021; Sanderson and Spacey, 2021) as well as in biomedical research, healthcare research and academic medical contexts (Athanasiou et al, 2016; Lewis et al, 2016; Valantine et al, 2016). It is suggested that, without such programmes, women and individuals from minority ethnic backgrounds may not be able to access suitable mentors. The difficulties of access to supportive individuals in the workplace may be amplified in environments in which there is cultural, racial or gender bias; exclusion; and high levels of competition. The structures that inhibit access to support are strong, and individuals' access to influential and experienced mentors can be compounded by complex issues. Such issues include the fear of being stereotyped that can impact on one's performance and aspirations ('stereotype threat', Steele and Aronson, 1995; Spencer et al, 2016); acceptance of other people's emotions, thoughts or feelings about one's 'inferiority' or 'otherness' to such an extent that they identify with these projections ('projective identification', Sandler, 2018); and self-doubt, despite evidence of achievement, about one's competence and ability leading to feelings of fraud and inadequacy ('imposterism', Clance and Imes, 1978). These issues are well recognised, particularly in the context of race (Ellis et al, 2020; Kinouani, 2021) and gender (Schmulian et al, 2020).

Research indicates that carefully designed mentoring schemes may offer valuable support for researchers transitioning to independence. For example, in the US, a randomised controlled trial in clinical and translational contexts with 283 mentors as participants, found that a structured mentoring curriculum improved the mentors' skills (Pfund et al, 2014). After receiving eight hours of training that focused on mentoring competencies, mentors and mentees reported positive changes in mentoring competencies. Interestingly, even mentors who already had considerable existing mentoring experience saw improvement (Pfund et al, 2014). The complete curriculum, evidence and further materials relating to mentor training are now freely available through the University of Wisconsin's Institute for Clinical and Translational Research.

In universities, mentoring is often implemented to address gender equality, but programmes that specifically provide mentorship for minority ethnic students and staff to nurture inclusion are still rare in the UK. This is despite real concerns about marginalisation of minority ethnic doctoral students and staff in academia and inadequate opportunities for career progression for minoritised individuals (Arday, 2021). One way to address this challenge is to encourage individuals to identify people within and outside of organisations who could provide career mentorship (Kinouani, 2021). However, this is not a panacea: 'the point here is not to create a false illusion that what we achieve at work is only within our control; it is not. Rather it is to have a plan, to be prepared psychologically and practically' (Kinouani, 2021, p. 128).

Diversity interventions that focus solely on remedying White male overrepresentation may implicitly and erroneously assume that members of different groups all share similar experiences of marginalisation. This homogenisation may overlook important differences, conceal experiences and reduce the possibility that advocacy can boost career prospects (Miriti, 2020). All these unintended consequences—and others beside—highlight the importance of building attention to intersectionality into intervention design. Doing so is likely to improve their effectiveness and reduce any potential for harm. Indeed, there are some positive evaluations of approaches that include intersectionality. For example, Mary Armstrong and Jasna Jovanovic (2017) showed that mentoring and networking programs in STEM (science, technology, engineering, mathematics) designed with built-in intersectionality (for example, for women of colour) were more effective than those framed as non-intersectional, such as programmes only for women or only for people of colour.

In the US, as part of the 'Building Up a Diverse Workforce for Biomedical Research Trial' aiming to compare mentorship programmes, Gretchen White and colleagues (2021) surveyed underrepresented early career researchers (ECRs) and faculty in relation to the COVID-19 pandemic and the anti-racism movement in the US. Participants reported both direct and indirect discrimination, feelings of isolation and psychological distress. The authors found that mentoring and support, especially with 'shared affinity' or 'common social identity' designed into the intervention (for example, background, functional area; and race, gender, gender identity, culture and ethnicity), had a positive protective effect on those impacted by racial injustice (White et al, 2021, p. 6).

However, even well-intentioned intersectional processes benefit from careful attention to avoid entrenchment of further disadvantage and inadvertent promotion of the deficit model. A useful example is provided by Christine Nittrouer and colleagues (2018), who offer an examination of gender and ethnicity distribution among biomedical science graduate students and staff in the context of US research labs. The authors found that female and minority ethnic (excluding Asian) students were paired with advisors who were not as successful in relation to publications and *h*-indices and, as a result, might have received inadequate levels of support, reproducing the disadvantage of the advisors. This aligns with previous research, which showed that poor outcomes go hand in hand with the lower support, and the insufficient mentoring female and minority ethnic graduates receive when they begin their careers in research-extensive institutions (Zambrana et al, 2015). It is important to recognise that the so-called 'homophily' in mentoring—the pairing of mentor-protégé sets of similar gender and ethnicity characteristics—has been recommended for EDI practitioners who design such programmes. However, this needs to be carefully curated to boost the potential benefits of such mentoring.

There is a further complexity in mentoring dyads (pairs) that can arise from personality traits; therefore, a careful approach to pairing is crucial. In the study interviews that we conducted, one participant shared her experience in a research institute:

'People have somebody—they might have a buddy or a mentor, but it's not very much oriented on diversity [...] There's a lot of neglect, especially in research, because the more technical your research part becomes, the personality types who do it change. And some are very cognitive and are very introvert and probably do not understand that somebody else needs some attention. Attention is a difficult thing [...] but people thrive on it.' (Participant 5, woman, EDI practitioner)

As far as the provision of mentoring schemes specifically for EDI practitioners and informal change agents is concerned, there is very little available research or information about good practice. This is surprising given that EDI staff, and researchers who voluntarily engage in this topic, are frequently at the forefront of organisational resistance to change (Ahmed, 2012). One possible approach to meet the needs of these individuals is to set up a community of practice, either at institutional or interinstitutional level, in which mentoring and peer-to-peer support is naturally built in (Thomson et al, 2021a). This approach will be presented later in the chapter. An example of practice oriented towards sharing knowledge of lived experience in a mentoring format—reverse mentoring—is presented next.

Reverse mentoring

So-called 'reverse mentoring' is often recommended as part of anti-racist change interventions in research and academic institutions (Grewal, 2022). This approach is indeed mentoring in reverse. In reverse mentoring, more experienced staff are mentored and guided by people who are earlier in their careers. Through this, more experienced staff are provided with a structure through which they gain insights, ideally bridging divisions that may exist by virtue of workplace hierarchy, contract arrangements or generation (Murphy, 2012). In the context of EDI, reverse mentoring is usually understood as a process in which the mentee does not have the lived experience of protected characteristics compared with the mentor. This relatively new form of mentoring is starting to show promise as a means of promoting individual-level anti-discriminatory practice and reducing implicit bias. In turn, such practice helps to support organisational culture change in gender and race equality (Clutterbuck and Ragins, 2002; Murphy, 2012). This method can contribute to the wider EDI agenda of incorporating various social categories, such as race and ethnicity, neurodiversity, disability and LGBTQIA+, as well as characteristics that are not protected by equality legislation, but nonetheless have an impact on advantage in the workplace, such as care responsibilities or socio-economic status (Agbalaya, 2021; Johnson, 2021). There are several advantages of taking the reverse mentoring approach. It 'reverses and disrupts traditional deficit model' by 'reposition[ing] the disadvantaged group as expert by experience' (Johnson, nd). At the same time, this approach acts as an intervention into skewed power structures that reproduce systematic disadvantage through individual-level interactions.

In 2020, Wellcome Sanger Institute and Wellcome Connecting Science piloted diverse reverse mentoring to facilitate engagement around building awareness about the underlying impact of racism and microaggressions on Black people and minority ethnic groups (Matimba and Dougherty, 2021). It was felt that a safe space within the organisation was needed to foster open dialogue about sensitive issues to drive positive organisational culture, a wider understanding of diverse experiences, and inclusive practice. This led to the establishment of the Race and Ethnicity Network (Matimba, 2023, personal email communication, 7 April). The mentoring programme connected two individuals from different career stages, gender and race backgrounds to engage in conversation about racism and the obstacles that Black people face in the workplace and their communities as well as celebrating positive interventions in the organisation. The exchange included sharing individual challenges as well as elements of career advice and guidance, introduction to career-enhancing networks, and professional development (Matimba, 2023, personal email communication, 7 April).

The WT also introduced the Reverse–Diverse mentoring scheme, involving individuals from staff diversity networks (race and equity; LGBTQ+; women; disability) mentoring the executive leadership team. The aim of the programme was to create an inclusive culture. This included raising awareness of, and helping to remove, some of the barriers both inside and outside of WT that can cause people to think that they do not belong in science and research (Agbalaya, 2021). This was part of a broader commitment at WT to anti-racism (Puvanendran, 2021). Similarly, ReMEDI—Reverse Mentoring for EDI partnership—was piloted by Derbyshire Healthcare NHS Foundation, its Black and Minority Ethnic Colleagues Network Trust, and the University of Nottingham. The intervention was built around concepts of resistance, illumination, solidarity and empowerment. Later, and in response to the COVID-19 pandemic, the approach also included trauma, turbulence and trouble (Johnson, 2021). The ReMEDI team used a critical pedagogy approach (Freire, 1970) to explore the intervention, including tools such as critical purposive conversations, observation and storytelling.

The Reverse–Diverse scheme at WT required meticulous planning to attend to possible risks and complexities. For example, review workshops with participants were arranged during the scheme to check the levels of comfort and progress. To create a safe peer support group, three mentors were assigned to the same mentee. The scheme successfully raised awareness of experiences of people from minoritised groups among the executive leadership team. Positive benefits included the opportunity for mentors and mentees to have open dialogue, peer-learning among mentors, improved communication and relationship-building skills, as well as gaining valuable insights into the executive leadership team (Agbalaya, 2021). Evaluations of ReMEDI indicated that the intervention was generally well received, particularly in relation to reaction and satisfaction, learning, and application and implementation. Nearly 95 per cent of participants rated the intervention as 'excellent' or 'good', and just under 90 per cent felt that they had gained confidence in engagement with EDI conversations. Over 60 per cent intended to stay in some form of contact with their mentoring partner; and 84 per cent rated their mentor/mentee match positively (Johnson, 2021). Positive evaluation of reverse mentoring in the context of medical school also showed that the scheme raised awareness among senior members of staff of the challenges faced by underrepresented medical students. The study also demonstrated that reverse mentoring sensitised senior staff to the institution's responsibility for nurturing EDI, and helped to challenge the prevalent deficit model in making sense of the underrepresented students' experience (Curtis et al, 2021).

There are some possible risks with reverse mentoring approaches, and care should be taken. For instance, although reverse mentoring aims to challenge power hierarchies, the approach exposes the mentor who is minoritised and potentially already marginalised to the possibility that they may experience a sense that their career could be damaged. They could experience triggering of past or current trauma, and the whole process requires emotional labour.

One of the study participants reflected on the practice of 'reciprocal' mentoring (involving reverse mentoring) in their institution:

'So we do have informal coffee mornings that will encourage the execs and the reciprocal mentors to attend. We do have formal supervision as well, which is separate to maybe raise any issues that have come up within some of these conversations. If anything be particularly triggering or someone's been dismissive about particular comments, then this is something that can be raised. It's interesting to note that those who, from the EDI perspective, have volunteered to be a reciprocal mentor/reverse mentor had to apply for those roles to determine whether or not they would be able to provide enough conversation and input information to be worthwhile. But that all the execs, it was mandatory for them. They didn't have a choice [...] which I think some are supportive of, but others maybe not so much.' (Participant 7, woman, EDI practitioner)

The participant's reflection serves as a reminder to consider how the decision is made about who can be the mentor and to consider whether scheme engagement is optional for mentees. Several questions arise here. First, is it ever appropriate and ethical to make a decision about whether someone's lived experience of marginalisation or disadvantage qualifies them to be a mentor? Second, who makes decisions about mentorship and eligibility and on what basis? Third, if mentee participation is either obligatory or expected, how best should mentees be identified and then supported, including if they are uncomfortable or unwilling to engage? Such complex questions need to be asked and considered during intervention. Moreover, a number of mechanisms should be in place before and during a scheme. These might include training, careful pairing, safeguarding awareness, reporting procedures, regular reviews and provision of access to emotional support. As this kind of mentoring provides benefit to those who are in more privileged positions, the labour undertaken by mentors should be recognised and rewarded appropriately and adequately.

Finally, reverse mentoring is a relatively new approach. More evaluation is needed to appraise its value and consequences. Crucially, a careful and reflexive approach to design and implementation is required in recognition of diverse contexts and the needs of different research organisations.

Communities of practice

The short-term nature of many positions and contracts in health and biomedical research has an impact on institutional memory about EDI work. As individuals move onto new roles or organisations, collective memories relating to most or least effective interventions can be lost if they are not recorded and shared. One way to retain memory is the community of practice (CoP) approach. CoPs are 'groups of people who share a concern, a set of problems, or a passion about a topic, and who deepen their knowledge and expertise in this area by interacting on an ongoing basis' (Wenger et al, 2002, p. 2). Communities of practice act as a 'storage container' for organisational memory and related learning (van den Brink, 2020). CoPs are well established in other settings—such as education, teaching and learning, health services, and business and management—but their use to address EDI in health and biomedical research contexts is relatively new. Their potential could be considerable.

CoPs may help EDI leads and stakeholders to build memory of diversity practices; may help to nurture the transfer and maintenance of established knowledge; and can foster individual agency and activism in a safe and supportive group of like-minded peers. The collective strength of a CoP can bolster credibility and legitimacy of equality goals, helping to bring about change. Community structures also nurture generosity of knowledge sharing and learning that can be transferred at different points in time, and also across different locations if CoPs are interorganisational (Thomson et al, 2021a). CoPs do not have to be large, but those that are interorganisational can have strategic reach. This can include engagement with policy makers and science funders to bring about change, as demonstrated by the San Francisco Declaration on Research Assessment (SF DORA), which brings together research funders as a CoP focusing on fair and responsible research assessment.

The CoP approach has recently gained traction in relation to the European Commission research funding and research performing organisations to promote gender equality through institutional gender equality plans. These plans aim to boost gender equality in career and development, leadership and decision making, but also in incorporating gender into research content through mainstreaming of sex and gender analysis. Examples of gender equality projects include Gender Time,[1] TARGET[2] and ACT on Gender,[3] which have been successfully implemented and for which evaluations are positive (Barnard et al, 2017; Palmén and Müller, 2022; Wroblewski and Palmén, 2022).

One of the TARGET project's beneficiaries, Regional Foundation for Biomedical Research (RFBR) in Italy, established a CoP that transcended its own organisational setting and included relevant external, but local, research and innovation stakeholders. Stakeholders included research centres, universities, hospitals, the scientific community who typically submit applications for RFBR funding, policy makers from the local region, and internal RFBR management and scientific committees. This inclusive CoP design nurtured a meaningful dialogue among the stakeholders and enabled to align research priorities and identify key strategic priority areas in biomedical research in the locality, while paying close attention to gender equality. Evaluation of other CoPs involved in the TARGET project emphasised that 'while individual agency and activism has been a key driving force for structural change, the support provided by the extended CoP has proved instrumental in making change happen' (Palmén and Caprile, 2022, p. 66).

Further examples of CoPs set up for gender equality efforts arise from the ACT on Gender project: including Life Science CoP (LifeSci), and Funding Organisations for Gender CoP (FORGEN). LifeSci CoP comprised members from European research centres and university faculties oriented towards life sciences. The main aim of LifeSci was to find practical interventions to change institutional culture and to improve gender equality, for example through exchanging good practice in research evaluation. Improvements in research practice were thought to be needed at the meso-level in the research community because of the mobility of researchers due to short-term contracts and employment arrangements. The CoP was found to foster a sense of strength, ability to share and empowerment:

> [I]t's a huge source of strength and people are realising that they do not have to reinvent the wheel. They can take best practices from other people, they can reuse activities [...] And that's really important because ... everybody likes to invent the wheel [...] and a lot of time and effort is wasted. (Reiland et al, 2022, p. 128)

FORGEN, led by Science Foundation Ireland, was joined by 17 science funders in nine European countries; the strategic role of research funders meant that effects could be cascaded into and across the research ecosystem. For instance, towards the end of the project, FORGEN collaborated with SF DORA to produce practice guidelines to help funders to adopt narrative CV (see Chapter 6) into practice as a means of reducing the potential for bias in the evaluation of applications for research funding (Fritch et al, 2021).

Creating a CoP within or across a number of institutions can empower involved stakeholders. However, there are some unintended consequences worthy of note. For instance, in settings where informal, grassroots communities are already formed, attempts to formalise and create official committees may undermine the preceding work and create new, unwelcome administrative burden. This issue is articulated by our interview study participant reflecting on attempts to formalise a pre-existing initiative:

> 'We're going through the process [of setting up a CoP] and I'm kind of taking with one hand and I'm pushing back with the other because they're trying to formalise it and I'm like, 'what are you trying to do to our group?' [...] a lot of the things that they talk about we do and

we do with no administrative overheads and the kind of stuff that they're talking about, there's administrative overhead [...] I'm happy to do it as part of my role but I'm not going to be doing it if someone wants me to take minutes and write agendas for every meeting. You're gonna have to pay someone to do that. This is my thing to support my colleagues and say we lift each other up. The better I am, the better my colleagues are and we all go up. [...] for communities that don't have it already set up, it's fantastic, and there will be a role for it.' (Participant 8, woman, EDI practitioner)

Stakeholders considering setting up a CoP need to reflect on such potential impacts if informal mechanisms are already in place. The advantages and disadvantages of formalisation need to be discussed, and the underlying reason for formalisation needs to be explored, considered and made clear to everyone if formalisation is needed. Not doing so may undermine previous work as well as future motivation and efforts.

CoPs also need strategies to include individuals who would not typically engage with EDI initiatives, but who are nevertheless important and whose involvement helps to nurture the EDI agenda. This is illustrated by another interview participant, who suggested that there is a risk that CoPs may be an example of an 'echo chamber' in which individuals only encounter information or opinions that resonate with their own:

'It's really encouraging to have conversations with like-minded individuals and with partners and to understand that this is something that others are hoping to try to achieve and it develops [into] a community of shared practice. [...] I think it's encouraging, but I think also we tend to speak into an echo chamber, don't we? Whereas those that are interested will engage and those that aren't interested won't engage.' (Participant 7, woman, EDI practitioner)

BOX 5.1: FURTHER EXAMPLES OF CoPs

Examples of the CoP approach provided in this chapter are focused on gender equality efforts. However, this collective intervention offers much promise in other areas of equality and inclusion work and institutional change. Communities have also been set up to converge work relating to race equality, disability and LGBTQIA+:

- anti-Racism CoP at the Center for Engaged Pedagogy at Barnard College of Columbia University, US
- anti-Racism CoP at the Psychological Professions Network in the UK
- Campus Compact—a higher education association in the US
- the Higher Education Community of Practice at the International Association of Accessibility Professionals focussed on disability
- LGBTQ+ in STEM (see Farrell et al, 2018)
- LGBTIQA+ Prisma Ciencia (prismaciencia.org) in Science, Technology and Innovation

This participant's words capture the importance of engagement with the wider research community and stakeholders, so that the CoP has reach beyond immediate members. Ideally, some members should be in positions that enable them to influence the research ecosystem more widely. CoPs should also be open to new members, while resistance from anywhere in the organisation can be welcomed as a means of developing productive dialogue about barriers to organisational change. Other practical considerations relate to the role of facilitators who organise the CoP. For instance, Jovana Mihajlović Trbovc and colleagues (2021) suggest that facilitators should be formally recognised, trained, financially compensated, and provided with opportunities to share knowledge, challenges and good practice with other facilitators. The authors also suggest that an allocated budget is crucial for sustainability. Although this focus on resourcing has its place, it is also important to recognise that effective CoPs may be grassroots developments, making use of existing resources and job roles (Mihajlović Trbovc et al, 2021). Moreover, there should be more than one facilitator nurturing community engagement to reduce the workload, or an arrangement whereby the facilitating role is rotated among members on a regular basis.

A noteworthy CoP, described as a coalition—Equality, Diversity and Inclusion in Science and Health (EDIS)[4]—is a community of research-funding, research-performing and research-oriented organisations that collaborate to drive change in EDI issues across the science and health research sector. This CoP draws its strengths from the foundational members—a research centre (the Francis Crick Institute), an independent health research funder (WT), and a commercial pharmaceutical and biotechnology partner (GlaxoSmithKline). It has around 30 members to date and ongoing new member engagement. The members subscribe to terms of reference that ask organisations to be active in EDI and fully support the coalition's objectives, which aim to influence the wider research ecosystem, leverage the collective voice for policy change, and contribute financial and human resources. In keeping with CoP good practice for equality projects (Barnard et al, 2017), EDIS requires member organisations to nominate a representative to contribute to regular meetings and activities, who should be someone with an influential position in the member organisation, such that they are able to nurture positive and significant systemic change. Core EDIS activities—knowledge and sharing good practice (for example, 'DAISY' diversity data collection guidelines; see Chapter 1), discussing challenges and celebrating successes in a safe and supportive environment—are fundamental CoP activities that nurture the community as well as inclusive practice in the research ecosystem.

Resources are available that provide step-by-step guidelines to the CoP approach (see Cambridge et al 2005; Wanger-Trayner[5]) and, once a CoP is set up, co-creation methods that nurture communities and action plans are curated for institutional change towards equality (see Thomson et al, 2021b for gender equality). This includes CoPs providing a helpful structure for facilitating 'mentoring circles', in which one mentor works with a group of mentees as well as mentees working with each other

(Darwin and Palmer, 2009). Such ideas and resources can help those in health and biomedical settings to consider whether and how a CoP might be a helpful approach to nurturing EDI. Futher examples of CoPs are presented in Box 5.1.

Allyship

In the context of EDI, a person who acts as an 'ally' for others supports colleagues who may be marginalised or disadvantaged. Support that an ally may consider includes work to seek to understand circumstances; to take action; to stand in unity; and reflecting and developing their own views and practice. History provides early examples of allyship: White people in the abolitionist movement, men supporting the women's suffrage movement, and 'straight' allies in early LGB movements. Recent examples of allyship include men who stand against misogyny and everyday sexism, White people who engage in anti-racism, 'active' bystanders who counter words and deeds of others, able-bodied individuals who challenge ableism, and support for LGBTQIA+ groups.

Although often understood as a matter of individual choice and action, allies have impact on institutions through voice, visibility, challenge, action and reflexive practice that improves rights and access to opportunities and equal outcomes. Allyship is inherently relational and can influence norms of behaviour in research citizenship, and the words used by allies are part of the action that they take.

Individuals who wish to be allies can face certain challenges. For instance, those who support groups that are marginalised, disadvantaged or stigmatised may themselves face marginalisation (Williams et al, 2022). This is, to an extent, covered as 'discrimination by association' in the Equality Act 2010. It is also important to consider whether allyship may have negative consequences for those that the activity purports to support and whether allyship might contribute to 'activist burnout', or in appropriation of power from those already at disadvantage (Gorski and Erakat, 2019). The term 'ally' and its performance can also reproduce and reinforce existing power dynamics and can also alienate those that it seeks to engage (Dabiri, 2021).

Another challenge is that allies may feel concern that they have never lived the experiences of disadvantaged colleagues to be able to legitimately voice support. An interview participant described his desire to be an ally and give voice while also experiencing feelings of concern that he did not have 'lived experience'. As a counterpoint, the participant approached the conundrum through the use of evidence from research as a means of influence:

> '[B]eing a White middle-aged British man [...] I think can be an advantage because if I can learn from others and communicate my best I can, from my perspective, what my understanding of the situation is, then that adds a different voice to what's going on. It's not necessarily an important voiçe and it's not necessarily a less important voice either, and it may [draw] more understanding and more interest in the area generally. [...] And sometimes it feels like it's a case of three steps forward and two steps back, and that's the worry there, and that's where I think again being someone without any obvious disadvantage myself can have a key role to play [...] in 'I can't do... because I can't authentically describe the lived experience of people with minority background'. What I can do [is] communicate research evidence as clear and broad a way as possible.' (Participant 3, man, researcher, EDI advocate)

Striving to be an ally is a complex endeavour that requires ongoing engagement enhanced by reflexivity. Reflexivity can include work to unlearn existing assumptions about the wider system; thinking about one's own position and role; accountability for one's own actions and a commitment to learning from criticism; and consideration about whether an approach was constructive, if so why and, if not, then why not? All individuals in the research ecosystem can engage in action with impact, but it is equally important to learn from the expertise of historically marginalised groups to better understand and then act on systems of inequality (Nixon, 2019). Allyship requires action and depends on continuous engagement with and work to improve one's own understanding of impacts of inequality. Drawing on the literature on allyship (Nixon, 2019; Carlson et al, 2020; Verma, 2022), we synthesise the following thought guide for individuals who aspire to be allies (Figure 5.1).

Importantly, we acknowledge Emma Dabiri's (2021) critique of allyship and her call to develop impactful 'coalitions' of people who share similar interests. The central argument Dabiri offers is that, rather than 'transferring' privilege, individuals should coalesce over commonalities, in which a critical mass can identify shared interests in relation to the same structural issues. A concrete example of a coalition that works in the spirit of 'shared interests' (EDIS) is presented earlier in this chapter in 'Communities of practice'.

EDI training

Research and academic organisations typically offer EDI-related training as part of induction and professional development programmes. Increasingly, such training is mandatory. EDI training may be provided as a generic curriculum, or in relation to particular operational processes, such as staff recruitment, diversity in teams, or line management. There are several approaches that underpin training offers. Some focus on enabling participants to improve their understanding of the equality legislation, with a focus on the protected characteristics and associated legal duties, governance, policy and other frameworks. Others focus on individual attitudes and related behaviours, for instance through recognition of, and ways to tackle, implicit bias and training about the bystander effect

Figure 5.1: Allyship behaviours

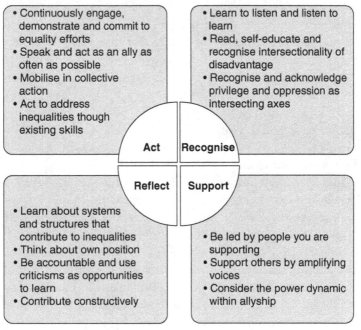

Source: Adapted from Nixon (2019), Carlson et al (2020) and Verma (2022).

and ways to become an active bystander. EDI training may use a variety of pedagogical approaches; for instance, awareness raising may comprise making lived experiences visible, giving voice and providing in-depth understanding, which can be both powerful and emotionally engaging.

Generic training or training about operational processes may be seen as more straightforward, all-encompassing, and efficient than more focused offers. Training that focuses on particular topic areas that relate to characteristics and inequalities—such as anti-racism or approaches to gender-based violence—may also be seen as more radical (Swan, 2009). Although such topic-specific training is rarer, there have been clear recommendations about their importance in research-related activities. Verma (2022) asserts that training is needed in relation to assessment of research quality and funding to prevent enactment of 'racial aggressions'.

While both types of training implicitly attend to awareness raising and educating individuals to guide micro-level interactions, practices and interrelations, they do not address the wider, systemic structures of inequality regimes (Acker, 2006). In a meta-analysis of diversity training provisions in organisational settings, over 60 per cent of evaluation studies prioritised individual-level cognitive and affective responses to diversity training. Just under 72 per cent of evaluated outcomes were based on self-reporting. These figures were even more polarised in training provisions among health service providers, where almost 87 per cent of evaluation studies prioritised individual-level responses, and 86 per cent were based on self-reporting (Devine and Ash, 2022). However, institutional inequalities that are built into policies and processes are not only often invisible to individuals but, even if they are obvious, they are outside the individuals' ability and reach to be addressed.

It has also been shown that individual-level diversity training may attract undesired resistance from its participants, who perceive such interventions with cynicism as a tick-box exercise, or preaching, or political correctness 'shoved down our throats' (Swan, 2009, p. 309), leading to backlash or resentment. The phenomenon of 'moral licensing' has also emerged to suggest that people tend to feel morally licensed towards misconduct, such as discrimination or exclusion, if they had done something 'good' (Monin and Miller, 2001; Uhlmann and Cohen, 2007), such as completing EDI training. This has prompted questioning whether 'attitude change through workshopping' is desirable and possible, since trainers and course designers eschew 'sustained or systematic study of the large body of literature and multiple perspectives on race, ethnicity or social antagonism [...] in favour of [...] quick fixes in the form of attitude adjustment' (Lasch-Quinn, 2002, pp. 166–167). Indeed, Patricia Devine and Tory Ash (2022) show that there are two issues that 'tarnish the silver bullet' of EDI training. First, limited and sometimes conflicting evidence that such programmes work or are effective in the long term for individuals taking part. Second, evidence about whether workplace environments become more equal, diverse and inclusive after such training programmes is inconclusive. These challenges are further unpacked by looking at how EDI training can be counterproductive. For instance, such training could activate rather than reduce stereotyping; could make people complacent by their believing that the organisation is responsible; could alienate the majority groups and cause resistance; and, lastly, training could create further resistance to mandatory and legislation-oriented initiatives if people feel compelled to undertake training that they would prefer not to attend (Dobbin and Kalev, 2018).

Moreover, although it is important to understand uptake of and response to EDI-related training to inform its design and planning, it is not always possible to know how a training offer will be received. Information about uptake can provide a view about the level of interest. As exemplified by one of the interview participants, training about equality and experiences can be popular compared with other courses:

'People say you know we want a talk on resilience, so we put on talks on resilience and nobody—well, half a dozen people—signed up, not enough to run the course. People say, we want a talk on mental health […] and people aren't turning up […] CV workshops […] But then I put on a talk about disability, what it's like for disabled colleagues to work [here] and we had 60 people turn up.' (Participant 8, woman, EDI practitioner)

Research organisations may need to develop tailored approaches to EDI training that sit alongside generic programmes. This should be done in reference to the recommendations on how to evaluate EDI training (Dobbin and Kalev, 2018; Davine and Ash, 2022) and adapted specifically to the context of research institutions that is presented in Box 5.2. Training should also

BOX 5.2: RECOMMENDATIONS FOR EDI TRAINING

Implementation of EDI training

- Assessing the status quo through EDI data to identify processes reproducing inequality and disadvantage, as part of targeted implementation design
- 'Blanket' training interventions supplemented with tailored provision targeting socially or processually connected individuals (for example, in skewed power dyads: applicant–reviewer; candidate–assessor; mentor–mentee; principal investigator–early career fellow, and so on)
- Regular and varied EDI training included in broader organisational/institutional initiatives, rather than a one-off (for example, embedded in organisational strategy)

Design of EDI training

- Build in opportunities to practise behaviours that increase empathy and emotional exchange through contact with the underrepresented groups
- The multicultural curriculum is framed as inclusive of the majority culture
- Introduce the 'moral licensing' concept to raise awareness among trainees
- Consider de-emphasising legislation-heavy content
- Consider voluntary enrolment, or give people a choice of various training options

Evaluation of EDI training

- Align the goals of EDI training with the evaluation outcomes. Avoid proxy measures (for example, overall satisfaction)
- Build in both behavioural and systems-level outcomes into evaluation depending on the context to monitor efficacy of EDI training
- Consider the uptake and no-show rates for evaluation and future planning

Monitoring of EDI training

- Measure diversity in award rates, success rates, hiring, persistence (students), retention (staff), perceived belonging, and perceived research culture among underrepresented groups, in a long-term pre/post evaluation model
- Monitor diversity in steering groups, review panels and promotion committees
- Monitor diversity and inclusion in collaborative research projects and authorship

Source: Adapted from Dobbin and Kalev (2018) and Davine and Ash (2022).

not be seen as a one-off, but a series of embedded opportunities to learn through group and self-learning, experiential learning, knowledge sharing, applying the lessons learned, and evaluation of the training.

Implicitly or unconsciously, we are biased

Attention to implicit (also known as unconscious) bias has increased in recent years. According to Google Ngram—which tracks the use of words and phrases in books—the frequency of 'unconscious/implicit bias' increased by around eight times from 2007 to 2019. Implicit bias is often understood as a cognitive process in which prejudice influences attitudes and behaviours. These result in a person showing favour to members of a particular group while marginalising members of another. Typically, this means favouring a more dominant or privileged group. The term 'unconscious' indicates that the person who holds biases is not aware of them. However, this term is often now thought to be unhelpful, as it could indicate that individuals should not be held responsible for any negative consequences stemming from their bias. In light of this concern, the term 'implicit' rather than 'unconscious' is often preferred in interventions that aim to address such bias.

Widely known about since its inception in 1998, Project Implicit sought to focus on bias by seeking 'to educate the public about bias and to provide a "virtual laboratory" for collecting data on the internet' (Project Implicit, 2011, np). Project Implicit is now a non-profit organisation consisting of support staff, consultants, scientists and an advisory board. The Implicit Association Tests (IAT)[6] developed by the project attracted millions of people to complete an online test to help them learn more about their biases, ranging from race, ethnicity and religion to age and body weight. Their popularity indicates encouraging levels of public interest in understanding their own personal bias. The tests have also been widely discussed and critiqued, including in relation to their ability to accurately predict behaviour and the clarity of the constructs that they measure, including whether the tests measure the construct of 'implicit' bias or something else (Bartels and Schoenrade, 2022). For EDI efforts, an unintended and unfortunate consequence of IATs is that they may normalise the existence of implicit bias: if we are all biased, what is the point of action, anyway?

At the same time as IAT focused attention on the subject, training in implicit bias has become a core component of EDI training in research environments. However, studies that assess the effectiveness of one-off implicit bias training sessions are inconclusive (Dobbin and Kalev, 2018) and it is not possible to claim that these interventions definitely work in all contexts and settings or, if they do, how long these effects last. So, why is this concept so popular? Laura Nelson and Kathrin Zippel (2021) suggest that implicit bias has become useful in academic science and related fields because the concept has five key features. First, implicit bias is seen as 'demonstrable', such that scientists claim they can prove its existence through experiments (such as the IATs). Second, it is 'relatable', because people have a good understanding of their own experience that they can refer to. Third, its applicability is 'versatile', as implicit bias can be used in various situations or intersections of social categories. Fourth, it is 'actionable', because implicit bias can be attended to through specific interventions (for example, focused training). Finally, implicit bias is considered to be 'impartial' and depoliticised, as the concept relegates the source of bias to a cognitive process, making it more impervious to resistance and 'palatable'. These five features have boosted the tractability of implicit bias interventions. At the same time, the features are also part of the limitations of implicit bias as a concept that can engender wider institutional change (Nelson and Zippel, 2021).

Finally, EDI practitioners may wish to be aware that—as with other EDI training—anti-bias programmes tend to emphasise that individuals rather than organisations should change. This focus on individuals means that there is limited potential for transformation of institutions and development of embedded practice. As such, anti–bias interventions should be complemented with wider system change approaches, which are discussed in the next chapter. However, this does not mean that bias should not be addressed: the five key features identified by Nelson and Zippel as needed to enable implicit bias to succeed are useful for any design of training curricula that nurture EDI.

Summary

- Good EDI practice and interventions at individual and community levels avoid the deficit model.
- There is benefit in considering how change and equality agents within organisations can make a difference to individuals who may be constrained by structural barriers and inequalities that have impacts within and outside organisational contexts.
- Both individual- and community-level interventions should be nurtured through collective and co-creative interventions that enable, empower and support disadvantaged and marginalised individuals as well as through amplifying the voices and lived experiences of minoritised groups and seeking shared interests.
- To underpin continuous improvement, it is important that reflection and evaluation are built into all EDI interventions.

SIX

How can EDI be nurtured through the research ecosystem?

This chapter focuses on organisational and system-based approaches to change. In Chapter 5, approaches that focus on nurturing EDI at individual and community levels were explored. This chapter presents strategies and interventions that can enable or hinder change at systems and organisational levels. As in the previous chapter, quotations from interview participants are included here for depth and to give voice to people's experiences of working within research organisations.

Research culture and assessment

Focus on research culture has recently gained momentum. Research culture 'encompasses the behaviours, values, expectations, attitudes, and norms of our research communities. It influences researchers' career paths and determines the way that research is conducted and communicated' (Royal Society, nd)—who is doing research, its integrity, diversity and inclusion. In health and biomedicine, key studies and reports describe researchers' experiences of the research cultures in which they work, with culture broadly understood as ways of working and beliefs or values held (for example, Wellcome Trust's work on inclusive culture; see Chapter 5 'reverse mentoring').

Organisations have responsibility for research culture, which impacts on individual wellbeing. Research environments can foster high levels of competition that can place individuals at a disadvantage in their careers. More positively, organisations and their leadership can commit to development of and support for culture that recognises and acts on inequalities to the benefit of everyone. EDI is immensely relevant to discussions of culture because the development of thriving, inclusive and vibrant research culture depends on work to enhance belonging and shared responsibility. Getting this work wrong, or failing to notice unintended consequences, could lead to benefit mainly to individuals who are already well connected, who can easily assimilate into dominant groups, and who already possess advantage.

Work on research culture has encompassed several key areas, including a focus on assessment of research at different points in the research lifecycle, ranging from grant applications (sometimes termed 'grant capture') to publications. Although this new focus takes place across all fields and disciplines, work to assess research quality is particularly visible in health and biomedical research, possibly partly because of the amount of funding available as well as the emphasis on published articles as outputs of interest and value.

There is still a need to assess whether interventions that address culture make a difference, including any impact on equality—ensuring that approaches that sit within the basket for any reforms do not create more adversity for researchers who may already be disadvantaged or marginalised. Among many, there are several interventions that relate to research culture and that address EDI in relation to assessment of research: attention to the use of publication metrics; inclusive considerations of one's academic age; voluntary disclosure of special circumstances in funding applications; and narrative curriculum vitae (NCV). These interventions will be presented in turn.

Attention to the use of publication metrics

How often a piece of research has been cited has been part of how we think about the importance of individual research studies, and their authors, for some time, with the first citation indices developed in the 1960s. Since then, the use of citation metrics has accelerated, aided by technological advances that change and speed up the ways that members of the research ecosystem seek and use research publications.[1] With the help of citation indices, journal metrics, and the h-index (citation count relating to individual researchers), metrics relating to publication have become incorporated into the evaluation of individuals and therefore into decisions about careers, including recruitment, promotion and performance evaluation. In recent years, shorthand methods to assess career potential and also individual success have included the name of the journal in which an individual has published, with high impact journals garnering prestige as well as individual h-indices.

These modes of assessment are fraught with problems. Metrics about journals (impact factors) are built on aggregate measures of a particular journal, which provide no detail about individual contributions. H-indices probably tell us more about the age of an author than the difference that their research has made to their field or the health of the public. Highly cited rankings are subject to a range of issues, including an absence of calibration by field and the impact of citation norms—including self-citation (van Noorden and Singh Chawla, 2019).

Several international initiatives provide system-wide approaches that can help reflect on, and refine or disrupt, the use of metrics in research assessment. These include the San Francisco Declaration on Research Assessment[2] (SF DORA), the Leiden Manifesto[3] (Hicks et al, 2015), the Metric Tide

Report (Wilsdon et al, 2015; Curry et al, 2022), and the Agreement on Reforming Research Assessment.[4] Within such initiatives, EDI is not always or necessarily explicit. However, at their heart are attempts to ensure that the research system appropriately recognises contributions to the system, that metrics are used responsibly, and that a broad range of career paths and perspectives are valued. EDI is interleaved between the pages of such work and there is recognition that research culture can pave the way for more equality.

Academic age: inclusive considerations

Some assessments of individuals' research are grounded in biological age. The use of age is subtle and often hard to describe, rather than direct and easy to identify and challenge. For instance, the length of time since the completion of a PhD is often used as a proxy for the level of experience and, as discussed above, the h-index is used as a measure of productivity. Bringing age into research assessment raises several challenges and issues that impact on equitable career pathways.

When publication rates are used as a measure of productivity, several assumptions are made visible. Peaks by age in such productivity vary across fields and disciplines. Insights into how age shapes a researcher's career have illuminated the changing, yet continuous, importance of researchers of all ages for research contributions. As such biological age is largely rejected as an indicator of research productivity and impact (Gingras et al, 2008).

In explorations of who receives research funding, biological age disadvantageously intersects with sex. For example, a study based in Quebec, Canada, found that women above the age of 38 years were awarded less research funding, produced fewer publications, and were disadvantaged in terms of citations of their work (Larivière et al, 2011). This may be due to individual, social, discipline-specific and institutional contexts. Alternative explanations may be various types of bias (confirmation bias, attribution error, solo status).

As individuals enter research careers at different points in their lives, the use of academic age attempts to counter the limitations of using biological age for individual-level bibliometrics and career trajectories. Academic age typically counts from the year of obtaining a PhD or the first publication, counted as year zero. Years since this point can be used as a criterion for a person's eligibility for a research grant or role. However, doing this may have unintended consequences for equity of outcomes, as it emphasises quantity and may cement the perception that 'more is better'.

The Faculty of Science at the University of Zurich[5] has developed a recommendation to address the potential that use of academic age may have an adverse and unwelcome impact. Applicants complete a one-page form (as well as provide their standard CV) with information about periods of time in which they were not working in a full-time capacity in research activities. The information can be used to allow decision makers to appreciate the 'academic age' of applicants more fairly (Petchey et al, nd). Another example is an initiative at the Swiss National Science Foundation (SNSF),[6] where applicants for funding are encouraged to state the duration of their work experience in research, after deducting time spent when either not working, disrupted, or undertaking other types of work. SNSF invites applicants to include clinical activities, parental-related reasons (maternity, paternity, adoption, parental leave, childcare), care duties, illness or accident, public service, training, education, part-time employment, non-scientific activities, or unemployment. This information allows evaluation of applicants' track record in relation to their net academic age. It is worth noting that in the SNSF scheme, the personal details are only reviewed by administrative officers rather than those who evaluate the applications for quality. The officers check the plausibility of the deductions, and evaluators only see the net academic age after the deductions have been checked. Further suggestions for the application of academic age include the use of age ranges, instead of the specific number of years, to discourage a focus on numbers, which may perpetuate disadvantage (Strinzel et al, 2021).

When information is collected about academic age, it is important that procedures are well developed and communicated. Otherwise, potential applicants may not wish to declare career breaks or other issues because of a fear of negative impact. One of the interview participants expressed this view:

> 'I think the negative thing rather comes from the applicants, that they are afraid if they put something in there, it has a negative impact. For example, we experimented a little bit with academic age calculation, trying to take off years of care and industry work, and whatever it was, illness, from the active research years, so that people [are] seen to be academically younger. And then they, applicants, were afraid that they seem to be too young, academically young, so they were afraid to give some information. [...] we still have a long way to go to educate ourselves and the panels how to deal with this information, what to do with it, how to take it into account. Because I think we all understood that we won't get the perfect number to compare one with the other.' (Participant 2, woman, EDI practitioner)

Participant 2 also highlighted that designing good practice is an ongoing process that requires much reflection and a close international dialogue among funders and research institutes.

Declaring 'special circumstances' in applications

Inviting researchers to declare special circumstances in funding applications allows evaluators to account for any pauses in, or changes to, research productivity. This is a relatively recent approach that seeks to account for circumstances so that the outcomes of decisions about research funding or careers are more equitable. A declaration of special circumstances relies on narrative self-declaration in which researchers may provide information about their individual context and circumstances to enable them to declare and explain how personal circumstances may have impacted adversely on their career and research. This approach might be appropriate when there is time to pay attention to such statements, and when the quantification of research age is not thought to be helpful or most relevant.

Asking individuals about circumstances has felt particularly relevant in light of the COVID-19 pandemic, in which restrictions on working environments (for example, lab closures) as well as changes to home circumstances (periods of lockdown) impacted on research productivity. These impacts are likely to be long-term, and more work is needed to consider how to address the inequalities that the pandemic amplified and highlighted. One of the interview study participants provided insight into how the impact of the pandemic was considered in her organisation, and how measures to address unequal impact remain work in progress in 2022:

'[W]e had the whole topic of especially with mothers and during these lockdown months [...] like all their struggles and their mental health problems. And of course, they came to talk to me and I was aware of the situation, how they worked night shifts to get the work done and everything was so overwhelming at home with the kids and the whole situation. So of course, this also influenced the work and like these acute measures we had to take to support especially these mothers so that they still can somehow maintain their mental health or maybe even physical health during the pandemic. [...] And now we're discussing if and how we could include this Covid time and self-declaration for example, people just give information about they could only work half time or they missed out three publications they couldn't hand in [...] to give room for these explanations. But we are not at the end of this process I would say. So, we're still discussing, trying, figuring out what is a good way to take away these disadvantages that some people experience. Because there were others who told me 'oh I could go to the countryside, I had a little cottage there and I could just do the writing all the time. Kids played in the garden'. So, it was really different for different people, and I think we have to find a measure to take away the disadvantages for the one group without giving too much advantage for the other group.' (Participant 2, woman, EDI practitioner)

Some of the challenges relating to special circumstances declarations in decisions about research funding are indicated in work by the Elizabeth Blackwell Institute for Health Research at the University of Bristol (where the authors are based). In 2019, just before the start of the COVID-19 pandemic, the Institute introduced a method to invite declarations of 'special circumstances' of funding applications to the Institute. Guidance to applicants asked them to focus on the impacts of any special circumstances rather than sharing detailed personal accounts. This was to protect applicants' privacy and to draw the attention of applicants and reviewers to the impact of circumstances on research rather than the nature of the circumstances. As the pandemic started and progressed, the impacts invariably related to the pandemic. Peer reviewers and panel members were provided with the statements, and applicants knew that their circumstances would be given to the evaluators. In 2022, the approach was evaluated, internally, with information collection looking at use of the invitation to declare special circumstances from 2019 to 2022. Over this period, one in four applicants made use of this section, with women more likely to declare special circumstances than men. Most applicants were satisfied with the opportunity to include this type of information even if when they did not need it and there was no evidence that those who declared special circumstances were disadvantaged by doing so. Importantly, the evaluation highlighted the need for further guidance for applicants and evaluators. The recommendations developed by the Institute focus on support for applicants (Figure 6.1) and guidance for evaluators (Figure 6.2).

Like many other approaches, asking researchers to describe the impact of special circumstances and then taking these into account during peer review or other forms of assessment has potential benefit, but also the potential for adverse unintended consequences. There is not a single method used across health and biomedical research. Moving forward, organisations might consider how best to invite a declaration of special circumstances, how to protect privacy, how to enable fair assessment, and to ensure that the approach fosters equity.

Narrative curriculum vitae

Similar to declarations of special circumstances, the narrative curriculum vitae (NCV) approach requires members of the research community to describe information about their research and careers. NCVs are a relatively new introduction to health and biomedicine and are part of a wider change across science and research. In an NCV, the individual provides information about their contribution across a number of domains. Depending on format, narrative information can complement lists such as publications, funding and committee membership; alternatively, narrative can replace those lists. The move to implement NCV stems from a concern that conventional scientific CV formats fail to enable individuals to demonstrate their contribution in the context of circumstances: 'A [standard] CV is unlikely to reflect the passion, perseverance, and creativity of individuals who struggled with limited resources and created their own opportunities for compelling research' (Schmid, 2013, np).

In the UK, an NCV format was developed jointly by the Royal Society and UK Research and Innovation (UKRI)

Figure 6.1: Special circumstances support for applicants

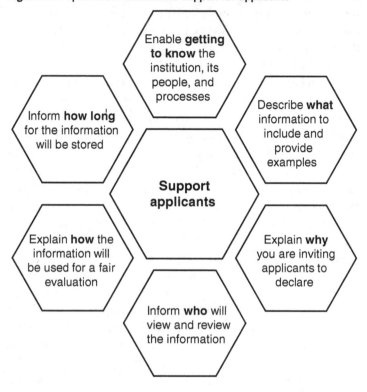

following a series of workshops in 2017 to encourage a conversation about research culture. A recommendation in the resulting report was to explore 'new forms of CVs that give a more rounded view of the individual' (Royal Society, 2019, p. 27). This led to the creation of the 'Résumé for Researchers' (R4R; see Box 6.1), which is starting to be implemented by some UK funders and organisations (for example, UKRI, University of Glasgow). Internationally, research funders have also experimented with various forms of NCV.

Consultations and evaluations of NCV are ongoing, with three important examples provided by the UK's National

Figure 6.2: Special circumstances guidance for evaluators

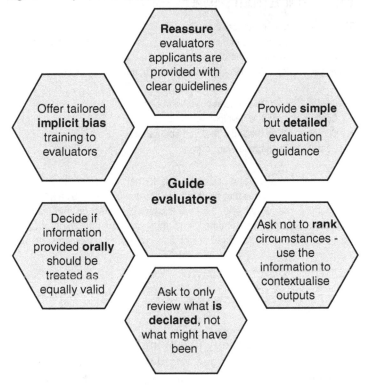

Institute for Health and Care Research (NIHR), the Luxembourg National Research Fund (FNR), and the SNSF.

First, NIHR conducted an extensive exploratory consultation combining data and materials from workshops, meetings, reports, survey feedback, review of existing CV alternatives, focus groups and interviews with research stakeholders. This was conducted before any decision to implement NCV. The resulting report offered recommendations for funders and a framework that indicates a range of considerations. Participants in the consultation recognised that 'implementing narrative CVs may offer some groups opportunities to apply to the

BOX 6.1: THE ROYAL SOCIETY'S RÉSUMÉ FOR RESEARCHERS (R4R)

The narrative format comprises distinct modules that signal valued areas and enable evaluators to compare qualitative information meaningfully. As well as standard indicators of research outputs (articles, papers, awards), the Royal Society's R4R asks for narratives structured into four modules:

- Contributions to the generation of knowledge
- Contributions to the development of individuals
- Contributions to the wider research community
- Contributions to broader society

Funders have been able to customise this format and decide in advance the kind of contributions they would like to see demonstrated by applicants to schemes. The format also invites disclosure of contextual information, such as career breaks, secondments, volunteering, part-time work, or time spent in different sectors that might have impacted on progression as a researcher.

NIHR and improve equity in funding […] in terms of career stage, diversity factors (age, gender, race, and disability), geographic affiliation, non-traditional research routes and career breaks' (Meadmore et al, 2022, p. 14). The work also highlights a need for training for and support to applicants and evaluators, a need to consider impact on certain groups such as people with dyslexia or dyspraxia and those for whom English is not their first language. Additionally, there might be the risk of inadvertent advantage for applicants who already frame their contributions with confidence. The report highlighted that the principles underpinning NCV were well understood, that work on implementation would take time and that there was a need to evaluate any impact of the NCV, including possible unintended consequences (Meadmore et al, 2022).

Second, while the NIHR's consultation explored NCV before any implementation, FNR evaluated NCV once implemented through a survey and across all fields of research. Although the evaluation was based on a small sample, and was limited in detail, the results were generally encouraging. Forty-eight per cent of applicants were satisfied with the NCV format, and 57 per cent agreed that the format allowed the researchers' achievements to be demonstrated and valued. Evaluators were even more positive: 71 per cent felt that the NCV format was useful, and only 9 per cent disagreed (FNR, 2022). Although the survey did not provide granular detail, such as information disaggregated by relevant individual characteristics, it provided useful signals about how the NCV is viewed.

Lastly, SNSF evaluated their 'SciCV'—a suite of CV components, including a newly introduced narrative element and academic age (presented earlier in this chapter), among other more conventional indicators. SciCV was designed to be used in assessment of grant applications. The evaluation of SciCV included a survey and interviews with applicants and reviewers of the CVs; text analysis of narrative aspects of the CVs and participant observation in research funding panel meetings. SNSF used disaggregated gender data (over 70 per cent were male), academic age, scientific fields within medicine and biology, and experience. The applicants and reviewers who tended to rate the narrative part and academic age as useful were described as more 'junior'. The authors of the evaluation suggested that this indicates that perhaps junior stakeholders were more open to the change. The findings from participant observation suggested that the new elements positively contributed to the overall picture of applicants, but did not challenge the reliance on 'traditional, publication-centred evaluation practices' (Strinzel et al, 2022, p. 1). The text analysis of the narratives showed no significant differences by gender; however, this should be treated with caution, as there were relatively few instances of the words that could indicate differences.

All three examples of consultation and evaluation indicate that there is a need to conduct and share further analysis of NCV. This arguably needs to include an examination of user experience alongside robust and long-term assessment of the impact of NCV adoption on careers, research and EDI. The following questions could frame an assessment of NCV implementation and should consider intersectionality:

- What are the experiences of diverse researchers and evaluators in terms of sex, gender identity, race, ethnicity, migrant status, class, disability, neurodivergence or institutional contexts, as well as discipline specificities?
- Does the implementation of the NCV enable recognition of a broad range of contributions of relevance to the role or funding opportunity?
- If the implementation of the NCV leads to recognition of a broad range of contributions, how does it contribute to systemic culture change?
- Do the pre-/post-evaluations of funding allocation indicate that the introduction of NCV has a positive impact on careers, research and EDI?
- How can information be sought about any unexpected consequences of NCV that have not yet been identified?
- What are the longer-term effects of NCV on EDI?
- How can EDI practitioners effectively collaborate across organisations to exchange good practice and evaluation assessment in ways that are as transparent as possible and as confidential as necessary?

In Figure 6.3, more granular questions are recommended. These are subsumed under a three-part reflexive cyclical framework. This framework encourages EDI practitioners to 'ask and listen', 'develop inclusive support for diverse groups', 'monitor equality effects and impacts', and then repeat the cycle, starting with feeding the evaluation and assessment back to the stakeholders, fine-tuning criteria and guidance, and monitoring impact.

Figure 6.3: Narrative CV development and evaluation cycle

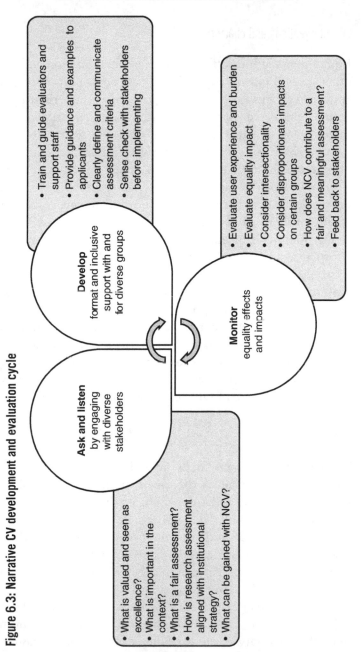

- Train and guide evaluators and support staff
- Provide guidance and examples to applicants
- Clearly define and communicate assessment criteria
- Sense check with stakeholders before implementing

Develop
format and inclusive support with and for diverse groups

- Evaluate user experience and burden
- Evaluate equality impact
- Consider intersectionality
- Consider disproportionate impacts on certain groups
- How does NCV contribute to a fair and meaningful assessment?
- Feed back to stakeholders

Monitor
equality affects and impacts

Ask and listen
by engaging with diverse stakeholders

- What is valued and seen as excellence?
- What is important in the context?
- What is a fair assessment?
- How is research assessment aligned with institutional strategy?
- What can be gained with NCV?

EDI interventions and charters

Equity interventions offer methods of precise action and are often termed either 'targeted' or 'positive' action. Targeted approaches focus on groups that are not characterised by a defined protected characteristic (for example, early career researchers). Positive approaches focus on groups whose members have a protected characteristic, and are disadvantaged, or have a specific need, or are underrepresented (for example, women or Black researchers). Charters, such as those focused on gender and race equality, can be instrumental in wider institutional change to the research ecosystem and culture by acting both as an incentive and recognition for institutional efforts in equality work. Athena SWAN, the Race Equality Charter, but also gender equality plans in the European Commission's research frameworks are mechanisms that exemplify institutional approaches to change and 'institutional isomorphism'. This occurs when an organisation desires to be in step with other organisations in the research ecosystem; or is compelled to demonstrate equality efforts for funding eligibility; or accepts and wishes to champion widely recognised and shared EDI norms and values (see Chapter 3).

Targeted interventions

Targeted interventions focus on redressing the impact of context and circumstances on careers. In the UK, fellowships for people who have had time away from their research careers provide good examples that address EDI. Such fellowships provide individuals with the opportunity to restart their research careers after breaks, for instance due to caring and family responsibilities. An example is provided by the Daphne Jackson Trust. The Trust was set up in 1992, in memory of the UK's first female professor of Physics. Daphne Jackson Fellowships are for anyone who has had a career break from research of at least two years for family, caring or health reasons.

Until 2003, all awardees were female (Times Higher Education, 2003); in 2020 the Trust broadened its remit from STEM (science, technology, engineering, mathematics) to all fields, and many fellowships have been held by individuals working in the fields of health and biomedicine as part of and related to STEM. Fellows are hosted in research organisations and provided with career development and support from the Trust. In an evaluation conducted by the Trust in 2022, the scheme was found to have demonstrable positive impact: awardees said that their fellowships improved confidence, enhanced prospects for future roles, and enabled them to restart research careers. Moreover, 90 per cent of fellows remained in research or teaching for at least five years after the fellowship, and over 70 per cent published a paper and continued to work in preferred role after the fellowship (Daphne Jackson Trust, 2022).

A further example of targeted intervention is achievement relative to opportunity (AR2O)—a principle that has been implemented in universities and research funders in Australia to guide evaluation of researchers. This intervention is consistent with the 'National Competitive Grants Program' mechanism implemented by the Australian Research Council—Research Opportunity and Performance Evidence[7] statement and assessment criterion—which aims 'to enable evaluation of a researcher's activities, outputs and achievements, in the context of career and life opportunities and experiences, including, where relevant, significant career interruptions' (Australian Research Council, nd, p. 2). AR2O sensitises evaluators to consider how researchers' careers are shaped by diversions and interruptions and other activities. By paying attention to non-linear careers, the approach is aligned with principles that also underpin NCVs and interventions to enable declarations of special circumstances. In relation to the evaluation of career progression, AR2O at Monash University stipulates that 'the overall quality and impact of achievements is given more weight than the quantity, rate or breadth of particular achievements relative to their personal, professional and other circumstances' and that consideration is given to 'the quantum or rate of productivity, the opportunity to participate in certain types of activities, and the consistency of activities or output over the period of consideration' (Monash University, nd, np). Like other initiatives in EDI, AR2O was seen as of particular relevance to women, with the approach described as 'essential to the equitable positioning of women in the academy' (Hill et al, 2014, p. 85), nurturing women's career and retention, and 'vital to modify practices that allow a reimaging of the ideal worker' (p. 106). Recently, Australia's National Health and Medical Research Council revised their AR2O guidance for peer reviewers to indicate that the impact of disability on research careers should be considered. This update followed an appeal made by Professor Justin Yerbury, a neuroscientist with motor neurone disease, who challenged a decision not to award funding due to his publication track record (Brock, 2021).

Positive action interventions

An international concept, 'positive action' is defined as 'consisting of proportionate measures undertaken with the purpose of achieving full and effective equality in practice for members of groups that are socially or economically disadvantaged, or otherwise face the consequences of past or present discrimination or disadvantage' (European Commission, 2009, p. 11). Law relating to positive action varies around the globe. This is a complex issue, and readers interested may wish to consult an international comparative report on positive action measures in the European Union, Canada, the US and South Africa (European Commission, 2009). In the UK, positive (affirmative) action is a lawful targeted intervention, relating specifically to people with protected characteristics under the Equality Act 2010.

In practice, in recruitment or promotion, the Equality Act allows that an individual from a disadvantaged or underrepresented group can be given selection preference provided that three conditions are met. First, the person cannot be less qualified or less suitable against the set criteria compared to another suitable candidate; second, the employer may not have a blanket policy of such favourable treatment; and, third, the positive action should be a balanced intervention that helps to lessen the disadvantage or increase the representation of this group (Equality Act 2010).

Positive action is different to positive discrimination. Positive discrimination does away with any conditions of special treatment. An example of this would be selecting a job applicant solely on the basis of their protected characteristic (for example, their race or sex), rather than another candidate who is better qualified for the role. This is currently not legal in the UK. Employers, however, can have a certain self-imposed target (for example, 30 per cent of professors to be female). Positive action schemes that aim to nurture a particular group of underrepresented researchers are direct interventions that can impact on bias and decision making. Such approaches can promote institutional change.

In the context of research, examples of positive action include bursaries and scholarships. Such interventions have been described as controversial but effective (Moody and Aldercotte, 2019), and therefore are often recommended to embed and 'embrace' EDI across the research ecosystem (Williams et al, 2019; Flinders, 2021, p. 3). Such tailored interventions, although complex to put into practice, are often thought to be of great value, as described by one of the EDI practitioners in the interview study:

'[W]e had from a private foundation money to fund women-only professorships. This application procedure was a little more difficult. It wasn't just women only, but we ended up with women only professorships. And this helped a lot to hire a group of women professors at one time and they didn't have to compete with other men. It was clear this position had to be 'we want women in this position'. And we had this selection procedure twice with three women professors for each round. So, we have six new women professors through this procedure. And this was like a huge … it's changed a lot because you really have this group and when you only have around 30 professors, then six means a lot.' (Participant 2, woman, EDI practitioner)

Talent and development programmes designed for people with certain protected characteristics are further examples of positive action. These take place within organisations when it is noted that employees who are members of particular groups are disadvantaged or underrepresented. For instance, one interview participant shared her experience of leading career programmes for women. She reflected on the process and the impact that the schemes had made:

'The Dean sends a letter every year for [the programmes for women]. He sends it to all the strategic department heads and the Dean asks […] would you please propose one of your talents. For the senior programme there is a selection, for the junior programme every department is allowed to propose one female talent. But the Dean does the request and tells the people it's important.

We're following the careers of the women as far as we can […] and you can see that the first group which was for the more senior women, started [over a decade ago] and 90 per cent of that first group is professor now.

We see the women who followed the more junior programme become more aware of what is needed in academia and their career chances grow.' (Participant 5, woman, EDI practitioner)

An example of a positive action intervention to address female underrepresentation in grant applications is offered by Science Foundation Ireland (SFI) (Fritch et al, 2019). SFI identified a gap between the percentage of applications from women

(25 per cent) to its early individual-level award (Starting Investigator Research Grant—SIRG) and the percentage of female research staff in STEM academic contracts at Irish universities at the time (39 per cent). SFI sought to address female underrepresentation through SIRG as a mechanism that would enable delivery of a positive action intervention. SFI put an incentive in place in the form of an increase in the number of allowed nominated candidates per institution from 10 to 12. However, this included a clause to ensure than there were no more than six male candidates nominated per institution. The submitted applications were treated equally regardless of the applicant's gender. As a result of the intervention, the percentage of applications from women jumped from the initial 25 per cent to 47 per cent, with the percentages of female awardees rising from 27 per cent to 50 per cent. SFI concluded that '[t]hese data support[ed] that there were suitable female candidates available, but that they were not being represented in the application pool' (Fritch et al, 2019, p. 197).

These examples have all been designed on the basis of consultation, reflection and evidence. It is also prudent to seek legal advice to be sure that the intervention does not inadvertently amount to unlawful discrimination. Any intervention requires full justification to demonstrate that it addresses either a disadvantage, a particular need, or low participation of the group with a particular protected characteristic. Moreover, such approaches need to be continually reviewed to monitor ongoing need to ensure that they remain fully justified. For example, if an intervention has been effective and has removed historic disadvantage, or addressed the particular need, or equalised the low participation, then the intervention may no longer be justified (Equality Challenge Unit, 2012). Unintended consequences may also arise if an organisation instigates a 'main' scheme and a 'positive action' scheme alongside one another, for instance, for research funding or staff recruitment. Such approaches could have negative impacts on engagement in main schemes. Observing legislation and remaining vigilant

BOX 6.2: POSITIVE ACTION INTERVENTIONS

Positive action is gaining momentum in health and biomedical sciences, with greater focus on initiatives that support people from minoritised ethnic backgrounds.

Sanger Excellence Fellowship

The Wellcome Sanger Institute, which conducts genomics research in the UK, launched a fellowship scheme in 2022 for early career researchers with an undergraduate degree and a PhD (or equivalent research experience) from a UK institution and from Black heritage backgrounds. The three-year fellowships provide salary, research consumables and a budget for training and conference attendance. The fellowship scheme is part of the Institute's work on race equity and the broader EDI programme of work, and seems in step with the increased awareness of underrepresentation of Black researchers in science and in scientific leadership roles. This is a long-term intervention with plans for the fellowship to be offered on an annual basis. In 2022, the Institute welcomed three fellows, including a joint fellowship with Cancer Research UK. The scheme has been expanded, with the Institute planning to award up to five fellowships in 2023 (Wellcome Sanger Institute, 2022).

Black British Heritage Summer School Scholarship

Another example of positive action in the context of biomedical science is implemented by Imperial College London. Launched in 2022, the scholarship covers summer school tuition fees for Black British Heritage undergraduate students.

---

to potential risks of positive action schemes is crucial to avoid inadvertently engaging in discrimination. As such, it is recommended to seek legal advice.

Equality charters

Equality charters in the UK, such as Athena Scientific Women's Academic Network (Athena SWAN, launched in

2005) and the Race Equality Charter (REC, launched in 2016) provide research and academic organisations with the opportunity to apply for recognition of their work on gender and race. The process for recognition requires submission of data and action plans that must address issues identified in the available quantitative and qualitative data. Institutions or schools and departments are able to apply for a gold, silver or bronze award, which respectively recognise the applying unit's level of commitment and action to promote gender equality. REC is currently available as an institutional-level award only. In health and biomedical research these two charters are widely recognised.

The overarching aim of the charters is to increase recruitment and promotion of women in fields with evidence of women's underrepresentation (through Athena SWAN); and to nurture institutions to 'identify and self-reflect on institutional and cultural barriers standing in the way of Black, Asian and Minority Ethnic staff and students' (through REC) (Advance HE, 2020, np). Athena SWAN has gained momentum in the UK (124 institutions with an award in 2022) as well as globally. The charter has also been adopted in Ireland in 2015, and inspired the establishment of similar initiatives in other countries, including Science in Australia Gender Equity (SAGE), STEM Equity Achievement (SEA) Change in the US, Dimensions in Canada; and the Gender Advancement for Transforming Institutions (GATI) in India.

In part, the prominence of Athena SWAN in health research can be traced to the relationship between NIHR funding and the Athena SWAN award level: between 2011 and 2020, NIHR required academic partners applying for some funding to have achieved a silver award of the Athena SWAN Charter. This was similar in spirit to a requirement imposed by the European Commission in 2022 for research organisations to have a gender equality plan or equivalent strategy in place to be able to apply for funding under its flagship Horizon framework (European Commission,

2021b). In 2020, when the requirement for a silver Athena SWAN award was removed, NIHR stated that it expected organisations to 'demonstrate commitment to tackling disadvantage and discrimination in respect of the nine protected characteristics set out in the Equality Act (2010)' (NIHR, 2020, np).

Responses to this change included concerns about deprioritising incentives that 'do drive action' against inequalities (Hewitt, cited in McIntyre, 2020), calls for 'reforming rather than rejecting' (Nature, 2020) and, more recently, making REC mandatory if racial inequalities are to be effectively addressed (Bhopal, 2022). Importantly, there is evidence that implementation of Athena SWAN action plans in medical sciences is linked to positive culture. For instance, a study by Pavel Ovseiko and colleagues (2019) on nearly 5,000 faculty and staff indicates the culture was more positive in medical sciences compared to social sciences. The authors attribute this difference to the extensive implementation of Athena SWAN action plans within medical sciences, linked to the NIHR funding eligibility criterion.

Athena SWAN has been evaluated in many studies, which have since produced good practice examples and recommendations. Key among these is work by Louise Caffrey and colleagues (2016), who explored how Athena SWAN was implemented in several departments in a university medical school that hosted a translational research organisation. The authors suggest that embedding Athena SWAN principles had a positive effect through creation of social spaces in which issues about inequalities and problematic practices could be openly discussed (Caffrey et al, 2016). However, engagement with charter processes required considerable amounts of time and resources for submission and to implement and monitor action plans. The study also indicated that such work is often disproportionately undertaken by women and minority ethnic individuals (Caffrey et al, 2016), while Holly Henderson and Kalwant Bhopal suggest that this work is delivered by staff 'qualified largely by lived experience as female academics rather than by expertise in these social systemic inequalities' (Henderson and Bhopal, 2021, p. 794). These efforts are considerable, and recognising that the work itself takes place within a broader context of social inequalities is a useful reminder of the ways in which the workload for EDI activities should be considered.

EDI practitioners also need to pay careful attention to structural factors to ensure that such equality efforts are successful. Rachel Palmén and Evanthia Kalpazidou Schmidt (2019) share their findings from analysis of case studies from the implementation of gender equality plans. The authors found that several factors had an impact on equality efforts. Some of the facilitating factors included ground-up participation and buy-in from all stakeholders and HR departments that boosted awareness and acceptance of equality efforts, and reduced resistance. Commitment from top levels of management was also crucial to leverage key change agents and ensure adequate allocation of resources. Further factors included a strategic framing of equality actions with other critical initiatives that can include 'framing of gender issues as inextricably linked to excellence in research' (2019, p. 4). This is highlighted through 19 case studies from Austria, Denmark, Germany, Hungary, Sweden and Spain within research and innovation in higher education, business enterprise and governmental sectors, including non-university research institutes.

Furthermore, reflecting on the effectiveness and lessons learnt from two EDI interventions—the Athena SWAN charter and the US National Science Foundation's ADVANCE programme—Sue Rosser and colleagues also highlight the importance of building a stronger focus on intersectionality beyond gender within such charters to advance their relevance; and inclusion of different types of institutions, such as non-profit organisations and scientific professional societies (Rosser et al, 2019).

Summary

- Many interventions relating to research culture also serve to nurture EDI, as they aspire to provision of 'equality to all'.
- The research ecosystem contains several EDI interventions that can be used by organisations as part of a comprehensive strategy.
- Reflection on EDI considerations—including attention to intersectionality—will assist organisations in the design and implementation of interventions and should reduce the risk of adverse unintended consequences that affect disadvantaged or underrepresented researchers.
- Positive action is an effective and recommended tool for nurturing EDI, although organisations implementing positive action should seek legal advice where appropriate to ensure that their approaches are likely to be lawful.
- Charters have a pivotal role to play in nurturing EDI through institutional change by means of rewarding, recognising and incentivising equality efforts. However, continuous evaluation is needed to monitor effectiveness, unintended consequences, and to ensure that charters remain relevant and fit for purpose.

Conclusion

This book provides evidence relating to inequalities in health and biomedical research careers and draws together information about EDI interventions. Writing the book was inspiring and fascinating. There was inspiration to be found in the narratives that people shared about their own work and experience, as well as in the numerous EDI initiatives designed and delivered across the research ecosystem and beyond. The aim was to provide as much material as possible that relates directly to health and biomedical research. However, many challenges that researchers face are shared with other research fields, and much can be learned from approaches beyond health and biomedicine, as well as from areas outside research. Drawing on concepts and literature from social sciences, human resource management, organisational theory and other fields has helped to provide foundations for a better understanding of EDI. Scholarship and evidence from broader studies of work and organisations provide key information about how the research ecosystem could progress.

Enhancement of equality, diversity and inclusion in careers enables research to deliver benefit more equally in local and global communities. The importance and ubiquity of health in people's everyday lives means that research matters and that, by attending to EDI, the health and biomedical research ecosystem can make a real difference within wider society as well as the ecosystem itself. The distinctive purpose of health and biomedicine usually represents a strong commitment to social justice through the focus on advancement of health. Ultimately, when the aim of a health and biomedical research organisation or ecosystem is to support the common good, there is a clear case that attending to diversity in the workforce meets both the business case and social justice rationale.

The book has provided an overview of key EDI concepts and issues, such as diversity characteristics, the ever-changing nature of the language of the concepts (Chapter 1), evidence of underrepresentation and bias (Chapter 2), rationales and approaches that can nurture diversity (Chapter 3), and the lived experience of marginalisation and exclusion (Chapter 4). To complement material focusing on challenges, the book contains a curated selection of EDI interventions that can be delivered through communities and individuals (Chapter 5) as well as schemes implemented across the research ecosystem to promote a wider system change (Chapter 6). It is worth noting that the selected possible interventions constitute 'work in progress', which research-performing and research-funding organisations co-create and update as a global and international community. As such, new pieces of evidence are likely to emerge to strengthen the case for the effectiveness of the selected interventions, and inevitably will demonstrate that certain interventions do not bring about the desired results. Rapid and much needed developments enabled by ongoing evaluation work disseminated within research and EDI groups can only be swiftly traced by well-networked and connected communities of practice (Chapter 6). Such communities can help to ensure EDI practitioners and advocates are not alone but feel part of a much larger and enthusiastic enterprise (Thomson et al, 2021a), and do not feel they have to 'reinvent the wheel' (van den Brink, 2020). Instead, they can cross-pollinate ideas, share knowledge and good practice, and nurture EDI work regardless of the different points that organisations and individuals may find themselves on during their EDI journeys.

The nature of EDI work means that its advocates and champions encounter resistance. The possibility of negative reactions to equality interventions should be recognised or even anticipated. Resistance and backlash against efforts to tackle inequalities may be individual and collective, formal and informal. In their work about resistance against gender equality, Michael Flood and colleagues suggest that reactions against progressive social change occur when individuals and groups resist, particularly those who, to some extent, enjoy or benefit from the status quo (2020). Reflecting on how resistance to equality has changed over time, Susan Faludi argues that, in the 1990s, backlash and resistance were largely invisible to the public eye and happening under the surface of civility. This has recently changed, such that these acts and attitudes are more noticeable, especially through social media and aided by political rhetoric (Faludi et al, 2020). Acknowledging that backlash is a 'predictable expression of the defence' of the status quo is an important part of realising social justice goals (Flood et al, 2020, p. 393). Moreover, examples can be shared to help to make the challenge visible to all, as highlighted recently by Joseph Hartland and Eva Larkai when describing experiences of hostility from media and social media sources to their work to decolonise a medical school curriculum (Hartland and Larkai, 2020).

EDI practitioners will themselves need strong and committed allies who are in positions of sufficient influence to catalyse change and champion EDI. At the same time, it is crucial to acknowledge that most—although not all—EDI-related work in research organisations is carried out by women and minorities, often unpaid and not always recognised for their stellar efforts on top of their daily tasks. Although this is changing as more organisations commit resources to EDI roles, much EDI work happens behind the scenes and is based on volunteering. In acknowledgement of this, we feel the deepest gratitude to the anonymous interview participants. They gave their time and energy to share generously their experiences with us and with readers of this book.

Before embarking on programmes attending to inequalities, EDI practitioners should ideally try to assess the status quo in their organisations. Organisations vary in the data available and cultural approaches to such information. Options may include collection or collation of quantitative information alongside qualitative data about experiences. Combining the two approaches provides balance and helps to understand reasons for the status quo as well as barriers and enablers to change. However, although such information is desirable, some absences in data should not discourage EDI practitioners from action. Data collection requires much time and resource, and care should be taken that this effort does not delay or distract practitioners from meaningful action based on the already available knowledge. Seeking dialogue with colleagues to gauge inclusion and possible organisational improvements can be an excellent way into EDI work and can support community-driven or organisational interventions that can foster change.

Our vision for the future centres on acceleration of EDI in health and biomedical research. The achievement of diversity in a workforce is one step along the journey, but a diverse workforce does not necessarily mean that an organisation's culture and related practice are inclusive. To support careers there is a need to nurture career progression, to consider who holds positions of influence, and to understand who has the chance to make decisions or take advantage of opportunities. To support careers in these and other ways requires systemic change that does not place responsibility solely onto individuals who may already be disadvantaged by the very system and structures that they are working within. Instead, all parts of the research ecosystem can embed EDI across the research lifecycle. This should mean that inclusion, diversity and equitable outcomes are normalised within research and are well within reach.

Appendix

Notes on study interviews included in the book

Direct quotations from individuals with experience of EDI matters are woven throughout the book. These are excerpts of semi-structured interviews that elicited views of members of the health and biomedical research community in relation to EDI work, interventions and support in their universities, institutes and research organisations. These complement and add voice to the representation of published evidence.

Interviews were conducted as a research study. To ensure that the plans for empirical research were examined before the work started, ethical approval was requested and received from the University of Bristol's Faculty of Health Science's Research Ethics Committee (approval number 9923, approval received in 2022). Invitation to participate was made through the authors' and the Elizabeth Blackwell Institute's professional networks and social media accounts (LinkedIn, Twitter). Twelve individuals offered to take part and eight semi-structured interviews were conducted by Dr Ola Thomson from May to July 2022. Four individuals decided not to take part before interview, either citing their busy work schedules or without giving a reason. All eight participants provided their informed consent, including agreement to audio-recording of interviews. After interviews and before publication, all participants have approved the inclusion of the excerpts in the book. At that point they were asked to provide their permission to publication and had the chance to edit the material or withdraw should they so wish.

In this process, one participant chose to edit the material, and all agreed to publication.

Five participants identified as women and three as men. Four participants were working in the UK, two in Spain, one in the Netherlands, and one in Germany. All worked in research or academic institutions and in departments focused on health and biomedicine. All participants actively engaged in some form of institutional change for gender equality, or all-encompassing equality and diversity, or both, through either formally assigned, paid, or informal roles (or sometimes both) in addition to their academic or research roles. Where appropriate, some details that could lead to identification have been altered in keeping with guidance and practice in qualitative research (Pope and Mays, 2020).

All eight interviews took place using online calling technology in the English language, and each lasted about one hour. Interview topic guides helped Dr Thomson to elicit background information about participants; information about their institution and its values, principles relating to EDI and known actions; and insights into wider issues relating to EDI, such as whether EDI matters in their research communities and how these issues are approached. Interviews were flexible, but all addressed the following research questions:

- How are institutional EDI interventions and support experienced and perceived by members of the health and biomedical research community?
- What are the experienced and perceived advantages and disadvantages of EDI institutional interventions and support, and to what extent do they progress and nurture individual researchers' careers and development?
- How has the COVID-19 pandemic shaped and influenced EDI interventions available to individuals within academic and research institutions?

Audio-recorded interviews were transcribed by a professional company under a contractual arrangement with the University of Bristol. Transcripts were read and annotated to highlight key areas of particular salience or relevance to the material in the book, and this work took place at the same time as ideas contained within the book were developed and drafted.

Rather than present material from the interviews as isolated 'findings' or 'results', the material is interleaved in the chapters. This presentation aims to illuminate and amplify voices of the EDI community who work tirelessly behind the scenes, and who grapple with the realities of equality work. Quotations from participants illustrate individual views of change agents, officers and stakeholders and to enrich our understanding of possible negative impacts on minoritised and disadvantaged researchers and the EDI community. This approach to knowledge creation and communication is rooted in feminist epistemologies (that is, studies of knowledge, for example, feminist standpoint theory) and methodologies that aim to provoke new ways of understanding and conveying material so generously provided by participants.

Notes

one

[1] https://www.gov.uk/definition-of-disability-under-equality-act-2010.

[2] Currently Title VII of the Civil Rights Act of 1964 prohibits discrimination not just because of one protected trait, but also because of the intersection of two or more protected bases: see: https://www.eeoc.gov/laws/guidance/section-15-race-and-color-discrimination#IVC.

two

[1] In the UK, APPGs are informal groups with no official status within Parliament, but can be a useful way to raise awareness of issues, which might later become topics for Select Committee inquiries, Parliamentary debates, or higher profile campaigns (British Science Association, 2021).

[2] https://www.nature.com/articles/d41586-021-03040-1.

[3] Australia, Brazil, Canada, China, India, Italy, Spain, UK and US.

[4] EURAXESS UK aids researchers in their career development, supporting mobility and acting as a support mechanism for researchers moving abroad or moving to the UK.

five

[1] https://gendertime.org/

[2] https://www.gendertarget.eu/

[3] https://act-on-gender.eu/

[4] https://edisgroup.org/

[5] https://www.wenger-trayner.com/

[6] https://implicit.harvard.edu/implicit/aboutus.html

six

[1] https://clarivate.com/webofsciencegroup/essays/in-memoriam-dr-eugene-garfield/

[2] https://sfdora.org/

[3] http://www.leidenmanifesto.org/

[4] https://coara.eu/

[5] https://www.mnf.uzh.ch/dam/jcr:12ac82ac-b2f0-46fd-bac6-5b2b853f9291/recruitingForExcellence.pdf

[6] https://www.snf.ch/media/en/Of9kzyITRoaTIIiN/SNSF_net-academic-age.pdf

[7] https://www.arc.gov.au/about-arc/program-policies/research-opportunity-and-performance-evidence-rope-statement

References

Acker, J. (2006) 'Inequality regimes gender, class, and race in organizations', *Gender & Society*, 20(4), 441. https://doi.org/10.1177/0891243206289499

Act for Autism (2022). https://actforautism.co.uk/home-2/

Acton, S. E., Bell, A. J., Toseland, C. P., and Twelvetrees, A. (2019) 'A survey of new PIs in the UK', *ELife*, 8. https://doi.org/10.7554/ELIFE.46827

Adébísí, F. I. (2019, 8 July) *The only accurate part of 'BAME' is the 'and'* .... https://folukeafrica.com/the-only-accepta ble-part-of-bame-is-the-and/

Advance HE (2020) *Race Equality Charter.* https://www.advance-he.ac.uk/equality-charters/race-equality-charter

Advance HE (2021a) *Equality in Higher Education: Staff Statistical Report 2021.*

Advance HE (2021b) *Equality in Higher Education: Student Statistical Report 2021.*

Advance HE (2022) *Guidance on the Collection of Diversity Monitoring Data: Updated Guidance on How to Collect Data About Personal Characteristics of Staff and Students in UK Higher Education.* https://s3.eu-west-2.amazonaws.com/assets.creode.advancehe-docum ent-manager/documents/advance-he/Advance%20HE%20Guida nce%20on%20the%20collection%20of%20diversity%20monitor ing%20data_1666776360.pdf

Agbalaya, T. (2021) *The Reverse Mentoring Revolution.* STEMM-CHANGE. https://www.stemm-change.co.uk/project-overv iew/the-reverse-mentoring-revolution-22-june-2021/

Ahmed, S. (2007) ' "You end up doing the document rather than doing the doing": Diversity, race equality and the politics of documentation', *Ethnic and Racial Studies*, 30(4), 590–609. https://doi.org/10.1080/01419870701356015

Ahmed, S. (2012) *On Being Included*. Duke University Press. https://doi.org/10.1215/9780822395324

Andersen, J.P., Wullum Nielsen, M., Simone, N.L., Resa E Lewiss, R.E. and Jagsi, R. (2020) 'Meta-Research: COVID-19 medical papers have fewer women first authors than expected', *eLife* 9: e58807. https://doi.org/10.7554/eLife.58807

Arday, J. (2018) 'Understanding mental health: what are the issues for Black and ethnic minority students at university?', *Social Sciences*, 7(196), 1–25. https://doi.org/10.3390/socsci7100196

Arday, J. (2021) 'Fighting the tide: understanding the difficulties facing Black, Asian and Minority Ethnic (BAME) doctoral students' pursuing a career in academia', *Educational Philosophy and Theory*, 53(10), 972–979. https://doi.org/10.1080/00131857.2020.1777640

Arday, J. (2022) 'No one can see me cry: understanding mental health issues for Black and minority ethnic staff in higher education', *Higher Education*, 83, 79–102. https://doi.org/10.1007/s10734-020-00636-w

Armstrong, M. A., and Jovanovic, J. (2017) 'The intersectional matrix: rethinking institutional change for URM women in STEM', *Journal of Diversity in Higher Education*, 10(3). https://doi.org/10.1037/dhe0000021

Ashe, S., Borkowska, M., and Nazroo, J. (2019) *Racism Ruins Lives: An Analysis of the 2016–2017 Trade Union Congress Racism at Work Survey*. University of Manchester. https://hummedia.manchester.ac.uk/institutes/code/research/projects/racism-at-work/tuc-full-report.pdf

Ashencaen Crabtree, S., Gatinao, A., Vasif, C., Nicholson, C., Boland, M. B., Speith, N., and Choe, J. (2017) *Talk About Success: BU Women Academics Speak*. Women's Academic Network.

Association of Graduate Careers Advisory Services (AGCAS) (2022) *What Do Graduates Do? 2021/22: Insights and Analysis from the UK's Largest Higher Education Survey.* https://graduatemarkettre nds.cdn.prismic.io/graduatemarkettrends/8d0f5a43-fe6e-4b78-b710-4c22baa1db5e_what-do-graduates-do-2021-22.pdf

Association of Medical Research Charities (2020) *Pandemic Threatens Future of Research as Early Career Scientists Look to Leave.* https://www.amrc.org.uk/news/pandemic-threatens-future-of-resea rch-as-early-career-scientists-look-to-leave#:~:text=Publis hed%3A%2028%20October%202020&text=The%20pande mic%20has%20had%20a,support%20for%20life%2Dsaving%20 discoveries

Athanasiou, T., Patel, V., Garas, G., Ashrafian, H., Hull, L., Sevdalis, N., Harding, S., Darzi, A., and Paroutis, S. (2016) 'Mentoring perception, scientific collaboration and research performance: is there a "gender gap" in academic medicine? An academic health science centre perspective', *Postgraduate Medical Journal*, 92(1092), 581–586. https://doi.org/10.1136/POSTGRADMEDJ-2016-134313

Australian Research Council (nd) 'Research Opportunity and Performance Evidence (ROPE) Statement', https://www.arc.gov. au/sites/default/files/2022-06/ROPE%20Statement.pdf

Bailyn, L. (2003) 'Academic careers and gender equity: lessons learned from MIT 1', *Gender, Work and Organization*, 10(2), 133–280. https://doi.org/10.1111/1468-0432.00008

Balloo, K., Hosein, A., Byrom, N., and Essau, C. A. (2022) 'Differences in mental health inequalities based on university attendance: intersectional multilevel analyses of individual heterogeneity and discriminatory accuracy', *SSM – Population Health*, 19. https://doi.org/10.1016/J.SSMPH.2022.101149

Banks, C. A., Gooberman-Hill, R., and Wainwright, D. (2020) 'An ethnography exploring the limits of dedifferentiation in the lives of adults with intellectual disabilities', 45(4), 344–354. https://doi.org/10.3109/13668250.2020.1799161

Bannerman, C., Guzman, N., Kumar, R., Nnebe, C., Setayesh, J., Venapally, A., and Sussman, J. H. (2020) 'Challenges and advice for MD/PhD applicants who are underrepresented in medicine', *Molecular Biology of the Cell*, 31(24), 2640–2643. https://doi.org/10.1091/mbc.E20-07-0444

Barnard, S., Hassan, T., Dainty, A., Polo, L., and Arrizabalaga, E. (2017). 'Using communities of practice to support the implementation of gender equality plans: lessons from a cross-national action research project'. In A.-S. Godfroy-Strauss and Y. Pouratt (Eds.), *Transferring, Implementing, Monitoring Gender Equality in Research Careers*. EDP Sciences.

Bartels, J. M., and Schoenrade, P. (2022) 'The implicit association test in introductory psychology textbooks: blind spot for controversy', *Psychology Learning and Teaching*, 21(2), 113–125. https://doi.org/10.1177/14757257211055200/FORMAT/EPUB

Baumer, N., and Frueh, J. (2021) *What Is Neurodiversity?* Harvard Health Publishing.

Beagan, B. L., Mohamed, T., Brooks, K., Waterfield, B., and Weinberg, M. (2021) 'Microaggressions experienced by LGBTQ academics in Canada: "just not fitting in … it does take a toll"', *International Journal of Qualitative Studies in Education*, 34(3), 197–212. https://doi.org/10.1080/09518398.2020.1735556

Bentley, A. R., Callier, S., and Rotimi, C. N. (2017) 'Diversity and inclusion in genomic research: why the uneven progress?', *Journal of Community Genetics*, 8(4), 255–266. https://doi.org/10.1007/S12687-017-0316-6/TABLES/1

Berlant, L., and Warner, M. (1998) 'Sex in public', *Critical Inquiry*, 24(2). https://doi.org/10.1086/448884

Bhopal, K. (2014) *The Experience of BME Academics in Higher Education: Aspirations in the Face of Inequality*. https://eprints.soton.ac.uk/364309/1/__soton.ac.uk_ude_personalfiles_users_kb4_mydocuments_Leadership%2520foundation%2520paper_Bhopal%2520stimuls%2520paper%2520final.pdf

Bhopal, K. (2020) 'Gender, ethnicity and career progression in UK higher education: a case study analysis', *Research Papers in Education*, 35(6), 706–721. https://doi.org/10.1080/02671 522.2019.1615118

Bhopal, K. (2022) '"We can talk the talk, but we're not allowed to walk the walk": the role of equality and diversity staff in higher education institutions in England', *Higher Education*, 1–15. https:// doi.org/10.1007/S10734-022-00835-7/METRICS

Botha, M. (2021) 'Academic, activist, or advocate? Angry, entangled, and emerging: a critical reflection on autism knowledge production', *Frontiers in Psychology*, 12, 4196. https://doi.org/ 10.3389/FPSYG.2021.727542

Bourne, C. (2020, 10 March) 'Is it time to legislate for dual discrimination?', *People Management*. https://www.peopleman agement.co.uk/experts/legal/is-it-time-to-legislate-for-dual-dis crimination#gref

Bowen, D. M. (2012) 'Visibly invisible: the burden of race and gender for female students of color striving for an academic career in the sciences'. In Gabriella Gutiérrez y Muhs, Yolanda Flores Niemann, Carmen G. González and Angela P. Harris (Eds.), *Presumed Incompetent: The Intersections of Race and Class for Women in Academia*. University Press of Colorado, Utah State University Press.

British Neuroscience Association (2020) *The Future of Neuroscience Research After COVID-19*. Survey report. https://www.bna.org. uk/media/resources/files/Covid-19_impact_neuroscience_resear ch_-_BNA_survey_results.pdf

British Science Association (2021) *Inquiry into Equity in the STEM Workforce*. https://www.britishscienceassociation.org/Handl ers/Download.ashx?IDMF=3d51130a-458b-4363-9b2b-d197a fc8382a

Brock, J. (2021, 19 January) '"Textbook case" of disability discrimination in grant applications', *Nature Index*. https://www. nature.com/nature-index/news-blog/textbook-case-of-disabil ity-discrimination-in-research-grant-applications

Buckner, E., Clerk, S., Marroquin, A., and Zhang, Y. (2020) 'Strategic benefits, symbolic commitments: how Canadian colleges and universities frame internationalization', *Canadian Journal of Higher Education*, 50(4). https://doi.org/10.47678/cjhe.vi0.188827

Buckner, E., Lumb, P., Jafarova, Z., Kang, P., Marroquin, A., and Zhang, Y. (2021) 'Diversity without race: how university internationalization strategies discuss international students', *Journal of International Students*, 11 (Special Issue 1). https://doi.org/10.32674/JIS.V11IS1.3842

Buldu, M. (2006) 'Young children's perceptions of scientists: a preliminary study', *Educational Research*, 48(1). https://doi.org/10.1080/00131880500498602

Butler, J. (1990) 'Gender trouble: feminism and the subversion of identity'. In J. Butler (Ed.), *Gender Trouble: Feminism and the Subversion of Identity*. Routledge. https://doi.org/10.4324/9780203824979

Byrd, M. Y., and Sparkman, T. E. (2022) 'Reconciling the business case and the social justice case for diversity: a model of human relations', *Human Resource Development Review*, 21(1). https://doi.org/10.1177/15344843211072356

Caffrey, L., Wyatt, D., Fudge, N., Mattingley, H., Williamson, C., and McKevitt, C. (2016) 'Gender equity programmes in academic medicine: a realist evaluation approach to Athena SWAN processes', *BMJ Open*, 6(9), e012090. https://doi.org/10.1136/BMJOPEN-2016-012090

Cambridge, D., Kaplan, S., and Suter, V. (2005) 'Community of practice design guide: a step-by-step guide for designing and cultivating communities of practice in higher education', *EDUCAUUSE*. https://library.educause.edu/resources/2005/1/community-of-practice-design-guide-a-stepbystep-guide-for-designing-cultivating-communities-of-practice-in-higher-education

Cardel, M. I., Dhurandhar, E., Yarar-Fisher, C., Foster, M., Hidalgo, B., Mcclure, L. A., Pagoto, S., Brown, N., Pekmezi, D., Sharafeldin, N., Willig, A. L., and Angelini, C. (2020) 'Turning chutes into ladders for women faculty: a review and roadmap for equity in academia', *Journal of Women's Health*, 29(5), 721–733. https://doi.org/10.1089/jwh.2019.8027

Careers Research & Advisory Centre (CRAC) Limited (2019) *Do Researchers' Early Careers Have to Be Precarious?* https://www. vitae.ac.uk/impact-and-evaluation/what-do-researchers-do/ vitae-wdrd-infographic-final-sept-19.pdf/@@download/ file/VITAE%20-%20WDRD%20Infographic%20Final%20S ept%2019.pdf

Careers Research & Advisory Centre (CRAC) Limited (2020) *Qualitative research on Barriers to Progression of Disabled Scientists: Report for the Royal Society.* https://royalsociety.org/-/media/policy/top ics/diversity-in-science/qualitative-research-on-barriers-to-prog ression-of-disabled-scientists.pdf

Careers Research & Advisory Centre (CRAC) Limited (2022) *What Do Researchers Do? Doctoral Graduate Employment, Activities and Earnings.* https://www.vitae.ac.uk/events/vitae-internatio nal-researcher-development-conference-2022/vitae-zone-2022/ wdrd-2022-report.pdf/@@download/file/What%20do%20rese archers%20do%20-%20docoral%20graduate%20employment,%20 activities%20&%20earnings%202022.pdf

Carlson, J., Leek, C., Casey, E., Tolman, R., and Allen, C. (2020) 'What's in a name? A synthesis of "allyship" elements from academic and activist literature', *Journal of Family Violence*, 35(8). https://doi.org/10.1007/s10896-019-00073-z

Cech, E. A. (2022) 'The intersectional privilege of white able-bodied heterosexual men in STEM', *Science Advances*, 8(24), eabo1558. https://doi.org/10.1126/SCIADV.ABO1558/SUPPL_FILE/SCI ADV.ABO1558_SM.PDF

Cha, S. E., and Roberts, L. M. (2019) 'Leveraging minority identities at work: an individual-level framework of the identity mobilization process', *Organization Science,* 30(4), 735–760. https://doi.org/10.1287/ORSC.2018.1272

Chambers, D., Preston, L., Topakas, A., de Saille, S., Salway, S., Booth, A., Dawson, J., and Wilsdon, J. (2017) *Review of Diversity and Inclusion Literature and an Evaluation of Methodologies and Metrics Relating to Health Research.* University of Sheffield.

Chambers, D. W. (1983) 'Stereotypic images of the scientist: the draw-a-scientist test', *Science Education*, 67(2). https://doi.org/10.1002/sce.3730670213

Clance, P. R., and Imes, S. A. (1978) 'The imposter phenomenon in high achieving women: dynamics and therapeutic intervention', *Psychotherapy: Theory, Research and Practice*, 15(3). https://doi.org/10.1037/h0086006

Clayton, S., and Opotow, S. (2003) 'Justice and identity: changing perspectives on what is fair', *Personality and Social Psychology Review*, 7(4), 298–310. https://doi.org/10.1207/S15327957PSPR0704_03

Clutterbuck, D., and Ragins, B. R. (2002) *Mentoring and Diversity: An International Perspective.* Butterworth-Heinemann.

Collins, P. H. (1990) 'Black feminist thought: knowledge, consciousness, and the politics of empowerment (perspectives on gender)', *Black Feminist Thought: Knowledge, Consciousness, and the Politics of Empowerment*, 138, 221–238.

Collins, P. H. (2015) 'Intersectionality's definitional dilemmas', *Annual Review of Sociology*, 41. https://doi.org/10.1146/annurev-soc-073014-112142

Cox, T. H., and Blake, S. (1991) 'Managing cultural diversity: implications for organizational competitiveness', *Academy of Management Perspectives*, 5(3), 45–56. https://doi.org/10.5465/AME.1991.4274465

Crear-Perry, J., Maybank, A., Keeys, M., Mitchell, N., and Godbolt, D. (2020) 'Moving towards anti-racist praxis in medicine', *The Lancet*, 396(10249), 451–453. https://doi.org/10.1016/S0140-6736(20)31543-9

Crenshaw, K. (1989) 'Demarginalizing the intersection of race and sex: a Black feminist critique of antidiscrimination doctrine, feminist theory and antiracist politics', *University of Chicago Legal Forum*, 1989(8). https://chicagounbound.uchicago.edu/uclf/vol1989/iss1/8

Curry, S., Gadd, E., and Wilsdon, J. (2022) *Harnessing the Metric Tide: Indicators, Infrastructures and Priorities for UK Responsible Research Assessment. Report of the Metric Tide Revisited Panel.* https://doi.org/10.6084/m9.figshare.21701624

Curtis, S., Mozley, H., Langford, C., Hartland, J., and Kelly, J. (2021) 'Challenging the deficit discourse in medical schools through reverse mentoring—using discourse analysis to explore staff perceptions of under-represented medical students', *BMJ Open*, 11(12), e054890. https://doi.org/10.1136/BMJOPEN-2021-054890

Dabiri, E. (2021) *What White People Can Do Next: From Allyship to Coalition.* Penguin Books.

Daphne Jackson Trust (2022) *Our Impact: Supporting Research Returners.* https://daphnejackson.org/wp-content/uploads/2022/05/Impact-Report_compressed-for-web.pdf

Dare, O., Jidong, D. E., and Premkumar, P. (2022) 'Conceptualising mental illness among university students of African, Caribbean and similar ethnic heritage in the United Kingdom', *Ethnicity & Health.* https://doi.org/10.1080/13557858.2022.2104817

Darko, N. (2021) *Engaging Black and Minority Ethnic Groups in Health Research. Hard to Reach? Demystifying the Misconceptions.* Policy Press.

Darwin, A., and Palmer, E. (2009) 'Mentoring circles in higher education', *Higher Education Research and Development,* 28(2), 125–136. https://doi.org/10.1080/07294360902725017

Department for Business Energy & Industrial Strategy (BEIS) (2021) *R&D People and Culture Strategy: People at the heart of R&D.* https://assets.publishing.service.gov.uk/government/uploads/system/uploads/attachment_data/file/1004685/r_d-people-culture-strategy.pdf

Department for Business Energy & Industrial Strategy (BEIS) (2022) *Bioscience and Health Technology Sector Statistics 2020.* https://www.gov.uk/government/statistics/bioscience-and-health-technology-sector-statistics-2020/bioscience-and-health-technology-sector-statistics-2020

Derrick, G. (2020) 'How COVID-19 lockdowns could lead to a kinder research culture', *Nature*, 581(7806), 107–108. https://doi.org/10.1038/D41586-020-01144-8

Derrick, G. E., Zimmermann, A., Greaves, H., Best, J., and Klavans, R. (2022) *Targeted, Actionable and Fair: Reviewer Reports as Feedback and Its Effect on ECR Career Choices*. https://doi.org/10.31235/OSF.IO/A8PSH

Devine, P. G., and Ash, T. L. (2022) 'Diversity training goals, limitations, and promise: a review of the multidisciplinary literature', *Annual Review of Psychology*, 73(1), 403–429. https://doi.org/10.1146/annurev-psych-060221-122215

Dias Lopes, A., and Wakeling, P. (2022) *Inequality in Early Career Research in the UK Life Sciences*. https://www.ukri.org/wp-content/uploads/2022/11/BBSRC-301122-BBSRCInequalityInEC RReport.pdf

DiMaggio, P. J., and Powell, W. W. (1983) 'The iron cage revisited: institutional isomorphism and collective rationality in organizational fields', *American Sociological Review*, 48(2). https://doi.org/10.2307/2095101

Department for Innovation, Universities and Skills (DIUS) (2008) *Innovation Nation. Department for Innovation, Universities & Skills Cm 7345*. https://assets.publishing.service.gov.uk/government/uploads/system/uploads/attachment_data/file/238751/7345.pdf

Dobbin, F., and Kalev, A. (2018) 'Why doesn't diversity training work? The challenge for industry and academia', *Anthropology Now*, 10(2), 48–55. https://doi.org/10.1080/19428200.2018.1493182

Drydakis, N. (2015) 'Sexual orientation discrimination in the United Kingdom's labour market: a field experiment', *Human Relations*, 68(11), 1769–1796. https://doi.org/10.1177/0018726715569855

Duda, R. (2015, August) 'Biomedical research', *80,000 Hours*. https://80000hours.org/career-reviews/biomedical-research/

Ellis, J., Otugo, O., Landry, A., and Landry, A. (2020) 'Interviewed while Black', *New England Journal of Medicine*, 383(25), 2401–2404. https://doi.org/10.1056/NEJMP2023999/SUPPL_FILE/NEJMP2023999_DISCLOSURES.PDF

Equality and Human Rights Commission (EHRC) (2019) *Racial Harassment in British Universities: Qualitative Research Findings.* https://www.equalityhumanrights.com/en/publication-download/racial-harassment-british-universities-qualitative-research-findings

Equality and Human Rights Commission (EHRC) (2022) *Public Sector Equality Duty.* https://www.equalityhumanrights.com/en/advice-and-guidance/public-sector-equality-duty

Equality Challenge Unit (2012) *Back to Knowledge Hub Guidance Notes Equality Act 2010: Positive Action Through Bursaries, Scholarships and Prizes.* https://s3.eu-west-2.amazonaws.com/assets.creode.advancehe-document-manager/documents/ecu/positive-action-through-bursaries-scholarships-and-prizes_1578925829.pdf

European Commission (2009) *International Perspectives on Positive Action Measures: A Comparative Analysis in the European Union, Canada, the United States and South Africa.* https://equineteurope.org/wp-content/uploads/2009/06/eul14058_def_en_090331_1__1.pdf

European Commission (2021a) *Roundtable on Equality Data.* https://commission.europa.eu/system/files/2021-12/roundtable-equality-data_post-event-report.pdf

European Commission (2021b) *Horizon Europe Guidance on Gender Equality Plans.* Publications Office of the European Union. https://doi.org/10.2777/876509

European Commission Directorate-General for Research and Innovation (2021c) *MORE4: Support Data Collection and Analysis Concerning Mobility Patterns and Career Paths of Researchers: Survey on Researchers in European Higher Education Institutions.* https://doi.org/10.2777/132356

European Institute for Gender Equality (EIGE) (2016a) *Gender Equality in Academia and Research—GEAR tool.* Publications Office for the European Union.

European Institute for Gender Equality (EIGE) (2016b) *Glossary & Thesaurus.* https://eige.europa.eu/taxonomy/term/1175

Faludi, S., Shames, S., Piscopo, J. M., and Walsh, D. M. (2020) 'A conversation with Susan Faludi on backlash, Trumpism, and #MeToo', *Signs: Journal of Women in Culture and Society*, 45(2), 336–345.

Fang, F. C., and Casadevall, A. (2016) 'Research funding: the case for a modified lottery', *mBio* 7(2). https://doi.org/10.1128/mBio.00422-16

Farrell, S., Chavela Guerra, R. C., Longo, A., and Tsanov, R. (2018) 'A virtual community of practice to promote LGBTQ inclusion in STEM: member perceptions and community outcomes', *2018 ASEE Annual Conference and Exposition, Conference Proceeding*. https://doi.org/10.18260/1-2--30028

Fatumo, S., Chikowore, T., Choudhury, A., Ayub, M., Martin, A. R., and Kuchenbaecker, K. (2022) 'A roadmap to increase diversity in genomic studies', *Nature Medicine*, 28(February), 243–250. https://doi.org/10.1038/s41591-021-01672-4

Flinders, M. (2021) 'Research leadership matters: agility, alignment, ambition: HEPI report 154', *Higher Education Policy Institute*. https://www.hepi.ac.uk/wp-content/uploads/2022/11/Research-Leadership-Matters-Agility-Alignment-Ambition.pdf

Flood, M., Dragiewicz, M., and Pease, B. (2020) *Resistance and Backlash to Gender Equality*. https://doi.org/10.1002/ajs4.137

Foster, D., and Scott, P. (2015) 'Nobody's responsibility: the precarious position of disabled employees in the UK workplace', *Industrial Relations Journal*, 46(4). https://doi.org/10.1111/irj.12107

Fredman, S. (2014) 'Addressing disparate impact: indirect discrimination and the public sector equality duty', *Industrial Law Journal*, 43(3), 349–363. https://doi.org/10.1093/INDLAW/DWU016

Freire, P. (1970) *Pedagogy of the Oppressed*. Seabury Press.

Frémeaux, S. (2020) 'A common good perspective on diversity', *Business Ethics Quarterly*, 30(2). https://doi.org/10.1017/beq.2019.37

Fritch, Rochelle, Hatch, Anna, Hazlett, Haley, and Vinkenburg, Claartje (2021) *Using Narrative CVs: Process Optimization and Bias Mitigation*. Zenodo. https://doi.org/10.5281/zenodo.5799414

Fritch, R., McIntosh, A., Stokes, N., and Boland, M. (2019) 'Practitioners' perspectives: a funder's experience of addressing gender balance in its portfolio of awards', *Interdisciplinary Science Reviews*, 44(2). https://doi.org/10.1080/03080188.2019.1603882

Gabriel, D. (2017) 'Introduction'. In D. Gabriel and S. A. Tate (Eds.), *Inside the Ivory Tower. Narratives of Women of Colour Surviving and Thriving in British Academia*. UCL Institute of Education Press.

Gabriel, D., and Tate, S. A. (2017) *Inside the Ivory Tower. Narratives of Women of Colour Surviving and Thriving in British Academia*. UCL Institute of Education Press.

Gibney, E. (2022) 'How UK science is failing Black researchers—in nine stark charts', *Nature*, 612(7940), 390–395. https://doi.org/10.1038/d41586-022-04386-w

Gingras, Y., Larivière, V., Macaluso, B., and Robitaille, J.-P. (2008) 'The effects of aging on researchers' publication and citation patterns', *PLoS ONE*, 3(12). e4048. https://doi.org/10.1371/journal.pone.0004048

Gloria, C. T., and Steinhardt, M. A. (2016) 'Relationships among positive emotions, coping, resilience and mental health', *Stress and Health*, 32(2), 145–156. https://doi.org/10.1002/SMI.2589

Gorski, P. C., and Erakat, N. (2019) 'Racism, whiteness, and burnout in antiracism movements: how white racial justice activists elevate burnout in racial justice activists of color in the United States', *Ethnicities*, 19(5). https://doi.org/10.1177/1468796819833871

Gotsis, G., and Kortezi, Z. (2013) 'Ethical paradigms as potential foundations of diversity management initiatives in business organizations', *Journal of Organizational Change Management*, 26(6). https://doi.org/10.1108/JOCM-11-2012-0183

Gould, J. (2015, 3 December) 'How to build a better PhD', *Nature*, 528(7580), 22–25. DOI: 10.1038/528022a

Government Equalities Office (2009, April) *Equality Bill: Assessing the Impact of a Multiple Discrimination Provision: A Discussion Document*. The National Archives. https://webarchive.nationalarchives.gov.uk/ukgwa/20110602110323/http://www.equalities.gov.uk/pdf/090422%20Multiple%20Discrimination%20Discussion%20Document%20Final%20Text.pdf

Grewal, M. K. (2022) 'Operations and processes'. In A. Verma (Ed.), *Anti-racism in Higher Education: An Action Guide for Change*. Policy Press.

Gunaratnam, Y. (2012) 'Researching race and ethnicity'. In Y. Gunaratnam (Ed.), *Researching Race and Ethnicity*. SAGE Publications Ltd. https://doi.org/10.4135/9780857024626

Gurbuz, E., Hanley, M., and Riby, D. M. (2019) 'University students with autism: the social and academic experiences of university in the UK', *Journal of Autism and Developmental Disorders*, 49(2). https://doi.org/10.1007/s10803-018-3741-4

Gurnett, J., and Morton, T. (2021) *Connecting and Communicating with Your Autistic Child: A Toolkit of Activities to Encourage Emotional Regulation and Social Development*. Jessica Kingsley Publishing.

Guyan, K. (2022) *Queer Data: Using Gender, Sex and Sexuality Data in Action*. Bloomsbury Publishing.

Guyan, K., and Douglas Oloyede, F. (2019) *Equality, Diversity and Inclusion in Research and Innovation: UK Review*. UK Research and Innovation.

Hafner-Burton, E., and Pollack, M. A. (2002) 'Mainstreaming gender in global governance', *European Journal of International Relations*, 8(3), 339–373. https://doi.org/10.1177/1354066102008003002

Hannam-Swain, S. (2017) 'The additional labour of a disabled PhD student', *Disability & Society*, 33(1), 138–142. https://doi.org/10.1080/09687599.2017.1375698

Hargrove, T. W. (2019) 'Light privilege? Skin tone stratification in health among African Americans', *Sociology of Race and Ethnicity*, 5(3), 370–387. https://doi.org/10.1177/2332649218793670

Hartland, J., and Larkai, E. (2020) 'Decolonising medical education and exploring White fragility', *BJGP Open*, 4(5), 1–3. https://doi.org/10.3399/BJGPO.2020.0147

Hartsock, N. C. M. (1998) 'The feminist standpoint revisited and other essays'. In N. C. M. Hartsock, *The Feminist Standpoint Revisited, and other Essays*. Routledge. https://doi.org/10.4324/9780429310881

Haynes, K., Metcalfe, J., and Yilmaz, M. (2016) *What Do Research Staff Do Next? 2016.* The Careers Research & Advisory Centre Limited.

Heffron, A. S., Braun, K. M., Allen-Savietta, C., Filut, A., Hanewall, C., Huttenlocher, A., Handelsman, J., and Carnes, M. (2021) 'Gender can influence student experiences in MD–PhD training', *Journal of Women's Health*, 30(1), 90–102. https://doi.org/10.1089/jwh.2019.8094

Henderson, H., and Bhopal, K. (2021) 'Narratives of academic staff involvement in Athena SWAN and race equality charter marks in UK higher education institutions', *Journal of Education Policy*, 37(5), 781–797. https://doi.org/10.1080/02680939.2021.1891576

Hewett, R., Douglas, G., and Mclinden, M. (2021) '"They were questioning whether I would even bother coming back". Exploring evidence of inequality in "access", "success" and "progression" in higher education for students with vision impairment', *Educational Review*, 75(2), 172–194. https://doi.org/10.1080/00131911.2021.1907315

Hicks, D., Wouters, P., Waltman, L., de Rijcke, S., and Rafols, I. (2015) 'Bibliometrics: the Leiden Manifesto for research metrics', *Nature*, 520(7548), 429–431. https://doi.org/10.1038/520429a

Higgins, M. C., and Dillon, J. R. (2007) 'Career patterns and organizational performance'. In H. Gunz and M. Peiperl (Eds.), *Handbook of Career Studies* (pp. 422–436). Sage Publications.

Higher Education Statistical Agency (HESA) (2021, 25 February) *Table 13: HE full-time academic staff by cost centre and contract salary: academic years 2014/15 to 2021/21.* https://www.hesa.ac.uk/data-and-analysis/staff/salaries

Higher Education Statistical Agency (HESA) (2022a, January) *Figure 17: HE qualifications obtained by CAH level 1 subject and sex: academic years 2019/20 to 2020/21.* https://www.hesa.ac.uk/data-and-analysis/sb262/figure-17

Higher Education Statistical Agency (HESA) (2022b, 17 February) *Chart 6: academic staff by sex, academic cost centre and cost centre group and academic year. Academic years 2014/15 to 2020/21.* https://www.hesa.ac.uk/data-and-analysis/staff/areas

Higher Education Statistics Agency (HESA) (2022c, 17 February) *Table 4: HE academic staff by ethnicity and academic employment function 2014/15 to 2020/21*. https://www.hesa.ac.uk/data-and-analysis/staff/table-4

Higher Education Statistical Agency (HESA) (2022d, 17 February) *Table 5: HE academic staff by disability and academic employment function: academic years 2020/21*. https://www.hesa.ac.uk/data-and-analysis/staff/working-in-he/characteristics

Higher Education Statistical Agency (HESA) (2022e, 17 February) *Table 42: UK domiciled qualifiers by subject of study and ethnicity 2019/20 to 2020/21*. https://www.hesa.ac.uk/data-and-analysis/students/table-42

Higher Education Statistical Agency (HESA) (2022f, 17 February) *Table 54: HE qualifiers by subject of study and domicile: CAH level 3*. https://www.hesa.ac.uk/data-and-analysis/students/outcomes#qualifiers

Hill, B., Secker, J., and Davidson, F. (2014) 'Achievement relative to opportunity: career hijacks in the academy', *Advances in Gender Research*, 19, 85–107. https://doi.org/10.1108/S1529-212620140000019004/FULL/XML

HM Government (2022) *Inclusive Britain: the Government's Response to the Commission on Race and Ethnic Disparities*. HM Government. https://www.gov.uk/government/publications/inclusive-britain-action-plan-government-response-to-the-commission-on-race-and-ethnic-disparities/inclusive-britain-government-response-to-the-commission-on-race-and-ethnic-disparities

Hollywood, A., McCarthy, D., Spencely, C. and Winstone, N. (2020) ' "Overwhelmed at first": the experience of career development in early career academics', *Journal of Further and Higher Education*, 44(7), 998–1012, https://doi.org/ 10.1080/0309877X.2019.1636213

Hull, L., Petrides, K. V., Allison, C., Smith, P., Baron-Cohen, S., Lai, M. C., and Mandy, W. (2017) ' "Putting on my best normal": social camouflaging in adults with autism spectrum conditions', *Journal of Autism and Developmental Disorders*, 47(8), 2519–2534. https://doi.org/10.1007/S10803-017-3166-5/TABLES/2

Hunt, L. (2021) *Diversity Data Collection: A Guide to Increasing Engagement and Response Rates*. Equality, Diversity and Inclusion in Science and Health (EDIS). https://edisgroup.org/wp-content/uploads/2021/05/Diversity-data-collection-engagement-tips-V1-1.pdf

Institute of Biomedical Science (2022) *What is Biomedical Science?* https://careers.ibms.org/discover-biomedical-science/what-is-biomedical-science/

Jackson, I. (2017) 'Inclusive ideas are not enough: academia does not empower Black women'. In D. Gabriel and S. A. Tate (Eds.), *Inside the Ivory Tower: Narratives of Women of Colour Surviving and Thriving in British Academia*. Trentham Books.

Jacobs, S.-E., Thomas, W., and Lang, S. (1997) *Two-Spirit People. Native American Gender Identity, Sexuality, and Spirituality*. University of Illinois Press.

Johnson, M. R. D., Bhopal, R. S., Ingleby, J. D., Gruer, L., and Petrova-Benedict, R. S. (2019) 'A glossary for the first World Congress on Migration, Ethnicity, Race and Health', *Public Health*, 172, 85–88. https://doi.org/10.1016/j.puhe.2019.05.001

Johnson, S. (nd) *The ReMEDI Project: Reverse Mentoring for Equality, Diversity and Inclusion (ReMEDI)*. The University of Nottingham. Retrieved 24 October 2022. https://onenhsfinance.nhs.uk/sites/default/files/blog-files/Reverse%20Mentoring%20The%20ReMEDI%20Project.pdf

Johnson, S. (2021) *The Reverse Mentoring Revolution*. STEMM-CHANGE. https://www.stemm-change.co.uk/project-overview/the-reverse-mentoring-revolution-22-june-2021/

Joice, W. and Tetlow, A. (2020) 'Baselines for Improving STEM Participation: Ethnicity STEM data for students and academic staff in higher education 2007/08 to 2018/19', *The Royal Society*. https://royalsociety.org/-/media/policy/Publications/2021/trends-ethnic-minorities-stem/Ethnicity-STEM-data-for-students-and-academic-staff-in-higher-education.pdf

Kapp, S. K., Gillespie-Lynch, K., Sherman, L. E., and Hutman, T. (2013) 'Deficit, difference, or both? Autism and neurodiversity', *Developmental Psychology*, 49(1), 59–71. https://doi.org/10.1037/A0028353

Kinouani, G. (2021) *Living While Black. The Essential Guide to Overcoming Racial Trauma.* Penguin Random House.

Klarsfeld, A. (2009) 'Managing diversity: the virtue of coercion'. In Mustafa F. Özbiligin (Ed.), *Equality, Diversity and Inclusion at Work: A Research Companion.* Edward Elgar Publishing. https://doi.org/10.4337/9781848449299.00036

Konrad, A. M. (2003) 'Special issue introduction: defining the domain of workplace diversity scholarship', *Group and Organization Management*, 28(1), 4–17. https://doi.org/10.1177/1059601102250013

Koo, K., and Nyunt, G. (2020) 'Culturally sensitive assessment of mental health for international students', *New Directions for Student Services*, (169), 43–52. https://doi.org/10.1002/SS.20343

Larivière, V., Vignola-Gagné, E., Villeneuve, C., Gélinas, P., and Gingras, Y. (2011) 'Sex differences in research funding, productivity and impact: an analysis of Québec university professors', *Scientometrics*, 87(3). https://doi.org/10.1007/s11192-011-0369-y

Larkworthy, C. (2022, 24 March) *World Autism Awareness Day: 2 April.* University of Oxford. https://staff.admin.ox.ac.uk/article/world-autism-awareness-day-2-april

Lasch-Quinn, E. (2002) *Race Experts How Racial Etiquette, Sensitivity Training, and New Age Therapy Hijacked the Civil Rights Revolution.* Rowman & Littlefield.

Lawson, A. (2022) *OSUN Inaugural Disability Justice Lecture: Disability, justice and the academy.* Open Society University Network. https://www.youtube.com/watch?v=yu2iSQhQoJY

Lawson, A., and Beckett, A. E. (2021) 'The social and human rights models of disability: towards a complementarity thesis', *International Journal of Human Rights*, 25(2), 348–379. https://doi.org/10.1080/13642987.2020.1783533

Lee, H. F., Miozzo, M., and Laredo, P. (2010) 'Career patterns and competences of PhDs in science and engineering in the knowledge economy: the case of graduates from a UK research-based university', *Research Policy*, 39(7). https://doi.org/10.1016/j.respol.2010.05.001

Levitt, H. M., and Ippolito, M. R. (2013) 'Being transgender: navigating minority stressors and developing authentic self-presentation', *Psychology of Women Quarterly*, 38(1). https://doi.org/10.1177/0361684313501644

Lewis, J. A., Mendenhall, R., Ojiemwen, A., Thomas, M., Riopelle, C., Harwood, S. A., and Huntt, M. B. (2021) 'Racial microaggressions and sense of belonging at a historically White university', *American Behavioral Scientist*, 65(8), 1049–1071. https://doi.org/10.1177/0002764219859613

Lewis, V., Martina, C. A., McDermott, M. P., Trief, P. M., Goodman, S. R., Morse, G. D., Laguardia, J. G., Sharp, D., and Ryan, R. M. (2016) 'A randomized controlled trial of mentoring interventions for underrepresented minorities', *Academic Medicine*, 91(7). https://doi.org/10.1097/ACM.0000000000001056

Lieff, S. J. (2009) 'Perspective: the missing link in academic career planning and development: pursuit of meaningful and aligned work', *Academic Medicine*, 84(10), 1383–1388. https://doi.org/10.1097/ACM.0B013E3181B6BD54

Lim, G. H. T., Sibanda, Z., and Erhabor, J. (2021) 'Eye-opener students' perceptions on race in medical education and healthcare', *Perspectives on Medical Education*, 10, 130–134. https://doi.org/10.1007/s40037-020-00645-6

Linton, M. J., Biddle, L., Bennet, J., Gunnell, D., Purdy, S., and Kidger, J. (2022) 'Barriers to students opting-in to universities notifying emergency contacts when serious mental health concerns emerge: a UK mixed methods analysis of policy preferences', *Journal of Affective Disorders Reports*, 7, 100289. https://doi.org/10.1016/J.JADR.2021.100289

Luxembourg National Research Fund (FNR) (2022) *Narrative CV: Implementation and Feedback Results.* https://storage.fnr.lu/index.php/s/YjunSGEQuSBRla8#pdfviewer

Mahayosnand, P. P., Zanders, L., Sabra, Z. M., Essa, S., Ahmed, S., Bermejo, D. M., Funmilayo, M., Sabra, D. M., and Ablay, S. (2021) 'E-mentoring female underrepresented public health student researchers: supporting a more diverse postpandemic workforce', *Health Security*, 19(S1), S72–S77. https://doi.org/10.1089/HS.2021.0042

Manfredi, S., Clayton-Hathway, K., and Cousens, E. (2019) 'Increasing gender diversity in higher education leadership: the role of executive search firms', *Social Sciences*, 8(6), 168. https://doi.org/10.3390/SOCSCI8060168

Manfredi, S., Vickers, L., and Clayton-Hathway, K. (2018) 'The public sector equality duty: enforcing equality rights through second-generation regulation', *Industrial Law Journal*, 47(3), 365–398. https://doi.org/10.1093/INDLAW/DWX022

Matimba, A., and Dougherty, M. (2021, 8 March) *Reverse Mentoring: Promoting Diversity and Positive Culture Change*. Wellcome Connecting Science. https://coursesandconferences.wellcomeconnectingscience.org/news_item/reverse-mentoring-promoting-diversity-and-positive-culture-change/

Maye, M., Boyd, B. A., Martínez-Pedraza, F., Halladay, A., Thurm, A., and Mandell, D. S. (2021) 'Biases, barriers, and possible solutions: steps towards addressing autism researchers under-engagement with racially, ethnically, and socioeconomically diverse communities', *Journal of Autism and Developmental Disorders*, 52(9). https://doi.org/10.1007/s10803-021-05250-y

McGettigan, A. (2013) *The Great University Gamble: Money, Markets and the Future of Higher Education*. PlutoPress.

McIntryre, F. (2020) 'Athena SWAN awards axed from NIHR funding requirement', *Research Professional*, 11 September 2020. https://www.researchprofessional.com/0/rr/news/uk/charities-and-societies/2020/9/Athena-SWAN-awards-axed-from-NIHR-funding-requirement.html

Meadmore, K., Recio-Saucedo, A., Blatch-Jones, A., Church, H., Cross, A., Fackrell, K., Thomas, S., and Tremain, E. (2022) 'Exploring the use of narrative CVs in the NIHR: a mixed method qualitative study', *NIHR Open Research*, 2(1). doi.org/10.3310/nihropenres.1115193.1

Mellors-Bourne, R., Jackson, C., and Hodges, V. (2012) *What Do Researchers Do? The Career Intentions of Doctoral Researchers 2012.* The Careers Research & Advisory Centre Limited.

Mihajlović Trbovc, J., Reiland, S., Reidl, S., and Damala, A. (2021) *ACT on Gender Policy Brief: How to Support Communities of Practice for Driving Institutional Change Towards Gender Equality* (pp. 1–5). ACT on Gender. https://www.act-on-gender.eu/sites/default/files/actongender_policy_brief_how_to_support_communities_of_practice_for_driving_institutional_change_towards.pdf

Milner, A., and Jumbe, S. (2020) 'Using the right words to address racial disparities in COVID-19', *The Lancet. Public Health*, 5(8), e419. https://doi.org/10.1016/S2468-2667(20)30162-6

Minello, A., Martucci, S., and Manzo, L. K. C. (2021) 'The pandemic and the academic mothers: present hardships and future perspectives', *European Societies*, 23(sup1), S82–S94. https://doi.org/10.1080/14616696.2020.1809690

Miriti, M. N. (2020) 'The elephant in the room: race and STEM diversity', *Bioscience*, 70(3). https://doi.org/10.1093/biosci/biz167

Molyneux, E., and Hunt, L. (2022) *Diversity and Inclusion Survey (DAISY) Question Guidance—Working Draft (v2).* Equality, Diversity and Inclusion in Science and Health (EDIS). https://edisgroup.org/wp-content/uploads/2022/05/DAISY-guidance-current-upated-May-2022-V2.pdf

Monash University (nd) *Achievement Relative to Opportunity.* Retrieved 1 December 2022. https://www.monash.edu/academicpromotion/achievement-relative-to-opportunity

Monin, B., and Miller, D. T. (2001) 'Moral credentials and the expression of prejudice', *Journal of Personality and Social Psychology*, 81(1). https://doi.org/10.1037/0022-3514.81.1.33

Moody, J., and Aldercotte, A. (2019) *Equality, Diversity and Inclusion in Research and Innovation: International Review*. UK Research and Innovation.

Mor Barak, M. E., Findler, L., and Wind, L. H. (2001) 'Diversity, inclusion, and commitment in organizations: international empirical explorations', *The Journal of Behavioral and Applied Management*, 2(2).

Mor Barak, M. E., Lizano, E. L., Kim, A., Duan, L., Rhee, M. K., Hsiao, H. Y., and Brimhall, K. C. (2016) 'The promise of diversity management for climate of inclusion: a state-of-the-art review and meta-analysis', *Human Service Organizations Management, Leadership and Governance*, 40(4). https://doi.org/10.1080/23303 131.2016.1138915

Moses, M. S. (2010) 'Moral and instrumental rationales for affirmative action in five national contexts', *Educational Researcher*, 39(3), 211–228. https://www-jstor-org.bris.idm.oclc.org/stable/27764 584?seq=8

Mulvey, M. R., West, R. M., Cotterill, L. A., Magee, C., Jones, D. E. J., Harris-Joseph, H., Thompson, P., and Hewison, J. (2022) 'Ten years of NIHR research training: who got an award? A retrospective cohort study', *BMJ Open*, 12, 46368. https://doi. org/10.1136/bmjopen-2020-046368

Murphy, W. M. (2012) 'Reverse mentoring at work: fostering cross-generational learning and developing millennial leaders', *Human Resource Management*, 51(4). https://doi.org/10.1002/hrm.21489

Nason, G., and Sangiuliano, M. (2020) *State of the Art Analysis: Mapping the Awarding Certification Landscape in Higher Education and Research*. Zenodo. https://doi.org/10.5281/ZENODO.4561664

National Association of Disabled Staff Networks (NADSN) (2021, 16 September) 'NADSN STEMM Recommendations Project: problem statement', *EDIS Members Annual Conference*.

National Center for Education Statistics (2021) *Digest of Education Statistics: Table 315.20*. https://nces.ed.gov/programs/digest/d21/ tables/dt21_315.20.asp

National Center for Science and Engineering Statistics (NCSES) (2020) *Postgraduation commitments of doctorate recipients: Table 42.* https://ncses.nsf.gov/pubs/nsf21308/table/42

National Institute for Health and Care Research (NIHR) (2020) *NIHR Responds to the Government's Call for Further Reduction in Bureaucracy with New Measures.* https://www.nihr.ac.uk/news/nihr-responds-to-the-governments-call-for-further-reduction-in-bureaucracy-with-new-measures/25633

Nature (2020) 'Equality and diversity efforts do not "burden" research—no matter what the UK government says', *Nature*, 586(7831), 643. https://doi.org/10.1038/D41586-020-03027-4

Nelson, L. K., and Zippel, K. (2021) 'From theory to practice and back: How the concept of implicit bias was implemented in academe, and what this means for gender theories of organizational change', *Gender & Society*, 35(3), 330–357. https://doi.org/10.1177/08912432211000335

Nittrouer, C., O'Brien, K. R., Hebl, M., Trump-Steele, R. C. E., Gardner, D. M., and Rodgers, J. (2018) 'The impact of biomedical students' ethnicity and gender', *Equality, Diversity and Inclusion*, 37(3). https://doi.org/10.1108/EDI-09-2017-0176

Nixon, S. A. (2019) 'The coin model of privilege and critical allyship: implications for health', *BMC Public Health*, 19(1), 1–13. https://doi.org/10.1186/S12889-019-7884-9/FIGURES/2

Noon, M. (2007) 'The fatal flaws of diversity and the business case for ethnic minorities', *Work, Employment and Society*, 21(4), 773–784. https://doi.org/10.1177/0950017007082886

Nuwer, R. (2020, 11 June) 'Meet the autistic scientists redefining autism', *The Scientist*. https://sites.udel.edu/seli-ud/files/2020/06/Meet-the-Autistic-Scientists-Redefining-Autism-Research-The-Scientist-Magazine®.pdf

Office for National Statistics (2009) *Measuring Sexual Identity: A Guide for Researchers.* Office for National Statistics.

Office for National Statistics (2022, 17 May) *A08: Labour Market Status of Disabled People*. Office for National Statistics . https://www.ons.gov.uk/employmentandlabourmarket/peopleinwork/employmentandemployeetypes/datasets/labourmarketstatusofdisabledpeoplea08

Oliver, M. (1983) 'Social work with disabled people'. In M. Oliver (Ed.), *Social Work with Disabled People*. Red Globe Press. https://doi.org/10.1007/978-1-349-86058-6

Oliver, M. (2013) 'The social model of disability: thirty years on', *Disability and Society*, 28(7). https://doi.org/10.1080/09687599.2013.818773

Oman, S., Rainford, J., and Stewart, H. (2015) 'Stories of access in higher education: a triumph (or failure) of hope over experience?', *Discover Society*, Dec(27). https://www.researchgate.net/publication/292604509_STORIES_OF_ACCESS_IN_HIGHER_EDUCATION_A_TRIUMPH_OR_FAILURE_OF_HOPE_OVER_EXPERIENCE

O'Neill, J., Tabish, H., Welch, V., Petticrew, M., Pottie, K., Clarke, M., Evans, T., Pardo Pardo, J., Waters, E., White, H., and Tugwell, P. (2014) 'Applying an equity lens to interventions: using PROGRESS ensures consideration of socially stratifying factors to illuminate inequities in health', *Journal of Clinical Epidemiology*, 67(1). https://doi.org/10.1016/j.jclinepi.2013.08.005

Organisation for Economic Co-operation and Development (OECD) (2015) *Frascati Manual 2015: Guidelines for Collecting and Reporting Data on Research and Experimental Development, The Measurement of Scientific, Technological and Innovation Activities*. OECD. https://doi.org/10.1787/9789264239012-en

Organisation for Economic Co-operation and Development (OECD) (2019) *Society at a Glance 2019*. OECD. https://doi.org/10.1787/soc_glance-2019-en

Organisation for Economic Co-operation and Development (OECD) (2021) 'Reducing the precarity of academic research careers', *OECD Science, Technology and Industry Policy Papers*, 113.

Ovseiko, P. V., Pololi, L. H., Edmunds, L. D., Civian, J. T., Daly, M., and Buchan, A. M. (2019) 'Creating a more supportive and inclusive university culture: a mixed-methods interdisciplinary comparative analysis of medical and social sciences at the University of Oxford', *Interdisciplinary Science Reviews*, 44(2), 166–191. https://doi.org/10.1080/03080188.2019.1603880

Özbilgin, M. F. (2009) 'Equality, diversity and inclusion at work: a research companion'. In M. F. Özbilgin (Ed.), *Equality, Diversity and Inclusion at Work: A Research Companion*. Edward Elgar Publishing. https://doi.org/10.4337/9781848449299

Ozturk, M. B. (2011) 'Sexual orientation discrimination: exploring the experiences of lesbian, gay and bisexual employees in Turkey', *Human Relations*, 64(8), 1099–1118. https://doi.org/10.1177/0018726710396249

Palmén, R., and Caprile, M. (2022) 'Relevance of a CoP for a reflexive gender equality policy: a structural change approach'. In R. Palmén and M. Caprile (Eds.), *Overcoming the Challenge of Structural Change in Research Organisations—A Reflexive Approach to Gender Equality* (pp. 53–69). Emerald Publishing Limited. https://doi.org/10.1108/978-1-80262-119-820221004

Palmén, R., and Müller, J. (Eds.) (2022) *A Community of Practice Approach to Improving Gender Equality in Research*. Routledge.

Palmén, R., and Schmidt, E. K. (2019) 'Analysing facilitating and hindering factors for implementing gender equality interventions in R&I: structures and processes', *Evaluation and Program Planning*, 77, 1–8. https://doi.org/10.1016/j.evalprogplan.2019.101726.

Parker, L., and Lynn, M. (2002) 'What's race got to do with it? Critical race theory's conflicts with and connections to qualitative research methodology and epistemology', *Qualitative Inquiry*, 8(1), 7–22.

Pelled, L. H., Ledford, G. E., and Mohrman, S. A. (1999) 'Demographic dissimilarity and workplace inclusion', *Journal of Management Studies*, 36(7). https://doi.org/10.1111/1467-6486.00168

Petchey, S., Gilland-Lutz, K., Siebert, P., and Kohler, I. (nd) *Recruiting for Excellence*. University of Zurich, Faculty of Science.

Pfund, C., House, S. C., Asquith, P., Fleming, M. F., Buhr, K. A., Burnham, E. L., Eichenberger Gilmore, J. M., Charles Huskins, W., McGee, R., Schurr, K., Shapiro, E. D., Spencer, K. C., and Sorkness, C. A. (2014) 'Training mentors of clinical and translational research scholars: a randomized controlled trial', *Academic Medicine: Journal of the Association of American Medical Colleges*, 89(5), 774–782. https://doi.org/10.1097/ACM.00000 00000000218

Pope, C., and Mays, N. (Eds.) (2020) *Qualitative Research in Health Care*, 4th edn. John Wiley and Sons.

Powell, M., and Johns, N. (2015) 'Realising the business case for diversity: a realist perspective on the British National Health Service', *Social Policy and Society*, 14(2), 161–173. https://doi.org/10.1017/S1474746414000025

Project Implicit (2011). https://implicit.harvard.edu/implicit/abou tus.html

Putnam-Walkerly, K., and Russell, E. (2016) 'What the heck does "equity" mean?', *Stanford Social Innovation Review*. https://doi.org/10.48558/YFPD-DE31

Puvanendran, K. (2021) 'How science-funding giant Wellcome is tackling racism', *Nature*, 594(7863). https://doi.org/10.1038/d41586-021-01582-y

Ramazanoglu, C., and Holland, J. (2002) *Feminist Methodology: Challenges and Choices*. Sage. http://search.ebscohost.com/login.aspx?dir ect=true&db=cat00012a&AN=bourne.506465&site=eds-live&scope=site

Ramnani, N. (2022) 'Ethnic minority participation in UKRI processes: evidence from six UKRI research councils over five years. DIV0093'. In *Written Evidence Submitted to the Diversity and Inclusion in STEM Inquiry*.

Rasekoala, E. (2022) 'Responsible science communication in Africa: rethinking drivers of policy, Afrocentricity and public engagement', *Journal of Science Communication*, 21(4), C01. https://doi.org/10.22323/2.21040301

Reiland, S., Palmén, R., and Kamlade, L. (2022) 'Disciplinary communities of practice for a greater gender equality in physics & life sciences'. In R. Palmén and J. Müller (Eds.), *A Community of Practice Approach to Improving Gender Equality in Research*. Routledge.

Rodó-Zárate, M. (2020) 'Gender, nation, and situated intersectionality: the case of Catalan pro-independence feminism', *Politics & Gender*, 16(2), 608–636. https://doi.org/10.1017/S1743923X19000035

Rosser, S. V., Barnard, S., Carnes, M., and Munir, F. (2019) 'Athena SWAN and ADVANCE: effectiveness and lessons learned', *The Lancet*, 393(10171), 604–608. https://doi.org/10.1016/S0140-6736(18)33213-6

Røttingen, J. A., Regmi, S., Eide, M., Young, A. J., Viergever, R. F., Ardal, C., Guzman, J., Edwards, D., Matlin, S. A., and Terry, R. F. (2013) 'Mapping of available health research and development data: what's there, what's missing, and what role is there for a global observatory?', *The Lancet*, 382(9900). https://doi.org/10.1016/S0140-6736(13)61046-6

Royal Society (2010) *The Scientific Century: Securing Our Future Prosperity*. Royal Society. https://royalsociety.org/~/media/Royal_Society_Content/policy/publications/2010/4294970126.pdf

Royal Society (2019) *Research Culture: Changing Expectations. Conference Report*. Royal Society. https://royalsociety.org/-/media/policy/projects/changing-expectations/changing-expectations-conference-report.pdf

Royal Society (nd) *Research Culture*. Royal Society. Retrieved 10 January 2023. https://royalsociety.org/topics-policy/projects/research-culture/

Ryan, W. (1971) *Blaming the victim*. New York: Random House.

Salem, V., Hirani, D., Lloyd, C., Regan, L., and Peters, C. J. (2022) 'Why are women still leaving academic medicine? A qualitative study within a London Medical School', *BMJ Open*, 12, 57847. https://doi.org/10.1136/bmjopen-2021-057847

Sanderson, R., and Spacey, R. (2021) 'Widening access to higher education for BAME students and students from lower socio-economic groups: a review of literature', *Journal of Higher Education Research IMPact*, 4(1).

Sandler, J. (2018) 'The concept of projective identification'. In J. Sandler (Ed.), *Projection, Identification, Projective Identification*. Routledge. https://doi.org/10.4324/9780429478574-2

Schiebinger, L. (1987) 'The history and philosophy of women in science: a review essay', *Signs: Journal of Women in Culture and Society*, 12(2). https://doi.org/10.1086/494323

Schiebinger, L., Klinge, I., Paik, H. Y., Sánchez de Madariaga, I., Schraudner, M., and Stefanick, M. (2020) *Gendered Innovations in Science, Health & Medicine, Engineering, and Environment.* European Commission. https://genderedinnovations.stanford.edu/Gendered%20Innovations.pdf

Schmid, S. L. (2013) 'Beyond CVs and impact factors: an employer's manifesto', *Science*. https://doi.org/10.1126/science.caredit.a1300186

Schmidt, J. (2016) 'Being "like a woman": Fa'afafine and Samoan masculinity', *Asia Pacific Journal of Athropology*, 17(3–4), 287–304. https://doi.org/10.1080/14442213.2016.1182208

Schmidt, J. M. (2017) 'Translating transgender: using Western discourses to understand Samoan fa'afāfine', *Sociology Compass*, 11(5). https://doi.org/10.1111/soc4.12485

Schmulian, D., Redgen, W., and Fleming, J. (2020) 'Impostor syndrome and compassion fatigue among postgraduate allied health students: a pilot study', *Focus on Health Professional Education: A Multi-Professional Journal*, 21(3). https://doi.org/10.11157/fohpe.v21i3.388

Scott, C. (2020) 'Managing and regulating commitments to equality, diversity and inclusion in higher education managing and regulating commitments to equality, diversity and inclusion in higher education', *Irish Educational Studies*, 39(2), 175–191. https://doi.org/10.1080/03323315.2020.1754879

Seeber, M., Cattaneo, M., Huisman, J., Paleari, S., and Be, M. S. (2016) 'Why do higher education institutions internationalize? An investigation of the multilevel determinants of internationalization rationales', *Higher Education*, 72, 685–702. https://doi.org/10.1007/s10734-015-9971-x

Showunmi, V., and Maylor, U. (2013) *Black Women Reflecting on Being Black in the Academy*. UCL Discovery. https://discovery.ucl.ac.uk/id/eprint/10012147/

Singer, J. (1998) *Odd People In: The Birth of Community Amongst People on the 'Autistic Spectrum'*. University of Technology.

Sivertsen, G. (2017) 'Unique, but still best practice? The Research Excellence Framework (REF) from an international perspective', *Palgrave Communications*, 3(1), 1–6. https://doi.org/10.1057/palcomms.2017.78

Spencer, S. J., Logel, C., and Davies, P. G. (2016) 'Stereotype threat', *Annual Review of Psychology*, 67. https://doi.org/10.1146/annurev-psych-073115-103235

Spier, J., and Natalier, K. (2023) 'Reasonable adjustments? Disabled research higher degree students' strategies for managing their candidature in an Australian university', *Disability & Society*, 38(8), 1365–1386. https://doi.org.10.1080/09687599.2021.1997718

Steele, C. M., and Aronson, J. (1995) 'Stereotype threat and the intellectual test performance of African Americans', *Journal of Personality and Social Psychology*, 69(5). https://doi.org/10.1037/0022-3514.69.5.797

Steinþórsdóttir, F. S., Brorsen Smidt, T., Pétursdóttir, G. M., Einarsdóttir, Þ., and le Feuvre, N. (2019) 'New managerialism in the academy: gender bias and precarity', *Gender, Work & Organization*, 26(2), 124–139. https://doi.org/10.1111/GWAO.12286

Stonewall (2022) *Global Workplace Equality Index*. https://www.stonewall.org.uk/global-workplace-equality-index

Strinzel, M., Brown, J., Kaltenbrunner, W., de Rijcke, S., and Hill, M. (2021) 'Ten ways to improve academic CVs for fairer research assessment', *Humanities and Social Sciences Communications*, 8(1), 1–4. https://doi.org/10.1057/s41599-021-00929-0

Strinzel, M., Kaltenbrunner, W., van der Weijden, I., von Arx, M., and Hill, M. (2022) 'SciCV, the Swiss National Science Foundation's new CV format', *BioRxiv*. https://doi.org/doi.org/10.1101/2022.03.16.484596

Sue, D. W., Capodilupo, C. M., Torino, G. C., Bucceri, J. M., Holder, A. M. B., Nadal, K. L., and Esquilin, M. (2007) 'Racial microaggressions in everyday life: implications for clinical practice', *American Psychologist*, 62(4). https://doi.org/10.1037/0003-066X.62.4.271

Swan, E. (2009) 'Putting words in our mouths: diversity training as heteroglossic organisational spaces'. In M. F. Özbilgin (Ed.), *Equality, Diversity and Inclusion at Work* (pp. 308–321). Edward Elgar Publishing.

Tate, S. A. (2017) 'How do you feel? "Well-being" as a deracinated strategic goal in UK universities'. In D. Gabriel and S. A. Tate (Eds.), *Inside the Ivory Tower: Narratives of Women of Colour Surviving and Thriving in British Academia* (pp. 54–66). Trentham Books.

Tate, S. A., and Bagguley, P. (2017) 'Building the anti-racist university: next steps', *Race Ethnicity and Education*, 20(3). https://doi.org/10.1080/13613324.2016.1260227

Thomson, A., Palmén, R., Reidl, S., Barnard, S., Beranek, S., Dainty, A. R. J., and Hassan, T. M. (2021a) 'Fostering collaborative approaches to gender equality interventions in higher education and research: the case of transnational and multi-institutional communities of practice', *Journal of Gender Studies*. https://doi.org/10.1080/09589236.2021.1935804

Thomson, A., Rabsch, K., Barnard, S., Dainty, A., Hassan, T. M., Bonder, G., Fernandez, B., and Romano, M. J. (2021b) *Community of Practice Co-creation Toolkit*, 2nd edn. The ACT Consortium. https://doi.org/10.5281/zenodo.5342489

Thomson, M. M., Zakaria, Z., and Radut-Taciu, R. (2019) 'Perceptions of scientists and stereotypes through the eyes of young school children', *Education Research International*. https://doi.org/10.1155/2019/6324704

Times Higher Education (2003, 24 October) 'Men who juggle campus and pampers', *Times Higher Education*. https://www.times highereducation.com/features/men-who-juggle-campus-and-pampers/184407.article

Times Higher Education (2021, 26 August) 'World university rankings 2022: methodology', *Times Higher Education*. https://www.timeshighereducation.com/world-university-rankings/world-university-rankings-2022-methodology#:~:text=Intern ational%20outlook%20(staff%2C%20students%2C%20resea rch)%3A%207.5%25&text=In%20the%20third%20internatio nal%20indicator,author%20and%20reward%20higher%20 volumes.

Truong, K. A., Museus, S. D., and Mcguire, K. M. (2016) 'Vicarious racism: a qualitative analysis of experiences with secondhand racism in graduate education', *International Journal of Qualitative Studies in Education,* 29(2). https://doi.org/10.1080/09518 398.2015.1023234

Uhlmann, E. L., and Cohen, G. L. (2007) ' "I think it, therefore it's true": effects of self-perceived objectivity on hiring discrimination', *Organizational Behavior and Human Decision Processes*, 104(2). https://doi.org/10.1016/j.obhdp.2007.07.001

UK Research and Innovation (UKRI) (2021a) *Detailed Ethnicity Analysis of Funding Applicants and Awardees: 2015–16 to 2019–20*. UKRI.

UK Research and Innovation (UKRI) (2021b) *Diversity Results for UKRI funding data: 2014–15 to 2019–20*. UKRI.

UK Research and Innovation (UKRI) (2022a) *Research Excellence Framework 2021: Results and Submissions*. UKRI. https://results2 021.ref.ac.uk/

UK Research and Innovation (UKRI) (2022b) *UKRI Glossary of EDI Terms*. UKRI. https://www.ukri.org/publications/ukri-gloss ary-of-edi-terms/ukri-glossary-of-edi-terms/

UK Research and Innovation (UKRI) (2022c) *Global Mobility of Research and Innovation: Personnel Evidence Report*. UKRI. http://www.discover.ukri.org/global-mobility-evidence-report-22

UK Research and Innovation (UKRI) (2022d) *UKRI Strategy 2022–2027: Transforming Tomorrow Together*. UKRI. https://www.ukri.org/publications/ukri-strategy-2022-to-2027/ukri-strategy-2022-to-2027/

Universities UK (UUK) (2022, 28 September) *Higher Education in Numbers*. Insights and Analysis. https://www.universitiesuk.ac.uk/latest/insights-and-analysis/higher-education-numbers

Universities UK (UUK) and National Union of Students (NUS) (2019) '*Black, Asian and Minority Ethnic Student Attainment at UK Universities: #Closingthegap*'. https://www.universitiesuk.ac.uk/sites/default/files/field/downloads/2021-07/bame-student-attainment.pdf

University College London (UCL) (2020) *Race Equality Charter Application Form*. University College London. https://www.ucl.ac.uk/equality-diversity-inclusion/sites/equality_diversity_inclusion/files/ucl-rec-submission-2020.pdf

Valantine, H. A., Lund, P. K., and Gammie, A. E. (2016) 'From the NIH: a systems approach to increasing the diversity of the biomedical research workforce', *CBE Life Sciences Education*, 15(3). https://doi.org/10.1187/cbe.16-03-0138

Valencia, R. R. (2010) 'Dismantling contemporary deficit thinking: educational thought and practice'. In R. R. Valencia (Ed.), *Dismantling Contemporary Deficit Thinking: Educational Thought and Practice*. Routledge. https://doi.org/10.4324/9780203853214

van den Brink, M. (2020) '"Reinventing the wheel over and over again". Organizational learning, memory and forgetting in doing diversity work', *Equality Diversity and Inclusion*, 39(4), 379–393. https://doi.org/10.1108/edi-10-2019-0249

van Dijk, H., van Engen, M., and Paauwe, J. (2012) 'Reframing the business case for diversity: a values and virtues perspective', *Journal of Business Ethics*, 111(1). https://doi.org/10.1007/s10551-012-1434-z

van Noorden, R., and Singh Chawla, D. (2019) 'Hundreds of extreme self-citing scientists revealed in new database', *Nature*, 572(7771), 578–579. https://doi.org/10.1038/D41586-019-02479-7

Verma, A. (2022) 'Research excellence assessments'. In A. Verma (Ed.), *Anti-racism in Higher Education* (pp. 90–95). Policy Press.

Viglione, G. (2020) 'Are women publishing less during the pandemic? Here's what the data say', *Nature*, 581(7809), 365–366. https://doi.org/10.1038/D41586-020-01294-9

Vinkenburg, C. J., Ossenkop, C., and Schiffbaenker, H. (2022) 'Selling science: optimizing the research funding evaluation and decision process', *An International Journal*, 41(9), 2040–7149. https://doi.org/10.1108/EDI-01-2021-0028

Wallace, S., Nazroo, J., and Bécares, L. (2016) 'Cumulative effect of racial discrimination on the mental health of ethnic minorities in the United Kingdom', *Public Health*, 106, 1294–1300. https://doi.org/10.2105/AJPH.2016.303121

Wellcome Sanger Institute (2022) *Sanger Excellence Fellowship*. https://www.sanger.ac.uk/about/equality-in-science/sanger-excellence-fellowship/

Wenger, E., McDermott, R. A., and Snyder, W. (2002) *Cultivating Communities of Practice: a Guide to Managing Knowledge*. Harvard Business School Press.

White, G. E., Proulx, C. N., Morone, N. E., Murrell, A. J., and Rubio, D. M. (2021) 'Recruiting underrepresented individuals in a double pandemic: lessons learned in a randomized control trial', *Translational Science*, 5, 185–186. https://doi.org/10.1017/cts.2021.843

Williams, M. T., Faber, S., Nepton, A., and Ching, T. H. W. (2022) 'Racial justice allyship requires civil courage: a behavioral prescription for moral growth and change', *American Psychologist*. https://doi.org/10.1037/amp0000940

Williams, P., Bath, S., Arday, J., and Lewis, C. (2019) *The Broken Pipeline: Barriers to Black PhD Students Accessing Research Council Funding*. https://leadingroutes.org/mdocs-posts/the-broken-pipeline-barriers-to-black-students-accessing-research-council-funding

Wilsdon, J., Allen, L., Belfiore, E., Campbell, P., Curry, S., Hill, S., Jones, R., Kain, R., Kerridge, S., Thelwall, M., Tinkler, J., Viney, I., Wouters, P., Hill, J., and Johnson, B. (2015) *The Metric Tide: Report of the Independent Review of the Role of Metrics in Research Assessment and Management.* ResearchGate. https://doi.org/10.13140/RG.2.1.4929.1363

Wilson, M. (2017) 'The search for that elusive sense of belonging, respect and visibility in academia'. In D. Gabriel and S. A. Tate (Eds.), *Inside the Ivory Tower: Narratives of Women of Colour Surviving and Thriving in British Academia* (pp. 108–123). Trentham Books.

Wong, S. H. M., Gishen, F., and Lokugamage, A. U. (2021) 'Decolonising the medical curriculum: humanising medicine through epistemic pluralism, cultural safety and critical consciousness', *London Review of Education*, 19(1). https://doi.org/10.14324/LRE.19.1.16

Woolston, C. (2021, 8 June) 'Researchers' career insecurity needs attention and reform now, says international coalition', *Nature*. https://www.nature.com/articles/d41586-021-01548-0

Woolston, C. (2022, 13 July) 'UK graduate students demand pay rise from nation's largest research funder', *Nature*. https://www.nature.com/articles/d41586-022-01934-2

World Health Organization (WHO) (2001) *International Classification of Functioning, Disability and Health: ICF.* World Health Organization. http://apps.who.int/iris/bitstream/handle/10665/42407/9241545429.pdf;jsessionid=F13928D0A3DA49F91C5FF721BC7030F1?sequence=1

World Health Organization (WHO) (2011) *World Report on Disability.* World Health Organization. https://www.who.int/publications/i/item/9789241564182

Wroblewski, A., and Palmén, R. (Eds.) (2022) *Overcoming the Challenge of Structural Change in Research Organisations: A Reflexive Approach to Gender Equality.* Emerald Publishing Limited.

Yang, Y., Tian, T. Y., Woodruff, T. K., Jones, B. F., and Uzzi, B. (2022) 'Gender-diverse teams produce more novel and higher-impact scientific ideas', *Proceedings of the National Academy of Sciences*, 119(36), e2200841119. https://doi.org/10.1073/PNAS.2200841119

Yerbury, J. J., and Yerbury, R. M. (2021) 'Disabled in academia: to be or not to be, that is the question', *Trends in Neurosciences*, 44(7), 507–509. https://doi.org/10.1016/J.TINS.2021.04.004

Yuval-Davis, N. (2015) 'Situated intersectionality and social inequality', *Raisons Politiques*, 58(2). https://doi.org/10.3917/rai.058.0091

Zambrana, R. E., Ray, R., Espino, M. M., Castro, C., Douthirt Cohen, B., and Eliason, J. (2015) ' "Don't leave us behind": the importance of mentoring for underrepresented minority faculty', *American Educational Research Journal*, 52(1), 40–72. https://doi.org/10.3102/0002831214563063

Zimmerman, A. M. (2018) 'Navigating the path to a biomedical science career', *PLoS ONE*, 13(9). https://doi.org/10.1371/journal.pone.0203783

Index

Page numbers in *italic* type refer to figures and tables.